REVOLUTION IN SCIENCE

Revolution in Science

How Galileo and Darwin Changed Our World

Mark L. Brake

palgrave
macmillan

First published in 2009 by PALGRAVE MACMILLAN® in the United States – a division of St. Martin's Press LLC, 175 Fifth Avenue, New York, NY 10010.

Where this book is distributed in the UK, Europe and the rest of the world, this is by Palgrave Macmillan, a division of Macmillan Publishers Limited, registered in England, company number 785998, of Houndmills, Basingstoke, Hampshire RG21 6XS.

PALGRAVE MACMILLAN is the global academic imprint of the above companies and has companies and representatives throughout the world.

Palgrave® and Macmillan® are registered trademarks in the United States, the United Kingdom, Europe and other countries.

ISBN: 978-0-230-20268-9

Library of Congress Cataloging-in-Publication Data is available from the Library of Congress.

A catalogue record of the book is available from the British Library.

Design by Macmillan Publishing Solutions.

First edition: December 2009

10 9 8 7 6 5 4 3 2 1

Printed in the United States of America.

For Been and Rosi,
who did not compel me to write this dedication

Contents

FIGURES

TABLE

INTRODUCTION

This is a tale of two revolutions. It is a story of history and adventure, science and invention, sex and absurdity, slavery and lunacy, murder and alchemy. A chronicle that sweeps continents and centuries, upending kings and cosmologies, religious dogma and the dark age of faith.

Bringing remarkable ages to vivid life, the narrative traces momentous events that twice turned the world upside down. A medieval revolution that shifted the Throne of God to the far reaches of the universe, and a Victorian revolution that struck at the heart of humanity itself.

Today we associate these respective revolutions with Galileo and Darwin. Each had a weapon of discovery: Galileo the telescope, Darwin the theory of evolution. With these weapons they helped prise open an alien universe, inhospitable to God and man alike.

Both Galileo and Darwin have become martyrs to science. Galileo's infamous trial at the hands of the Inquisition gave enormous prestige to the new experimental science. The drama of Darwinism was used to justify ruthless exploitation, the conquest of innocent peoples, and even war itself.

But our story is far bigger than just Galileo and Darwin.

For it has universal, as well as human, themes. Astronomy and evolutionary biology are historical sciences. They try to explain the sweep of phenomenal changes in Earth and Sky over the vastness of deep space and time.

Our study of these subjects must center on the evolving link between industry, empire, and the structure of scientific revolutions. Indeed, the very source of the word "revolution" as meaning social and political upheaval, originates from the days of Copernicus and Galileo. And the revolutions play witness to the original paradigm shifts. They are crucial not just to science but equally to society and culture.

Furthermore, our story will trace the origin and development of the modern malaise in the human condition, associated with man's separation from nature. And it allows us to contemplate the great lessons of science: how it works, how it errs, how it evolves.

So, this book tackles the topic in the most intelligent way, the long view, the evolutionary view. In science, more than in any other human tradition, we must look to the past to understand the now, and divine the future.

THE PLAN OF THE BOOK

Each revolution has a backstory.

The tale of our changing perception of the natural world is one that began in ancient times—the grandeur that was the ancient Greek world, the days of the Great Alexander. To a large extent, the intellectual and cultural brilliance of the Greeks has a profound effect on our narrative. So Part I of the book, "Wisdom of the Ancients," looks at the way in which the classical world influenced succeeding ages, especially in relation to the natural sciences of astronomy and biology.

Movements of economy and society, politics and people, play out on a global scale. Chapter 1, "Greek Sky: The World Before the Telescope," looks at origins. We look at the first application of mathematics to the human experience by the Pythagorean Brotherhood, the founders of what the world understands today as science. If it is the classical Greek world we have to thank for the conscious and unbroken thread of rational thinking, why *then*? Why in ancient Greece does all this develop? What were the conditions of culture and economy that led to such a remarkable history? And why was a member of the brotherhood murdered for the sake of mathematics?

The war of worldviews played out between Galileo and the Inquisition, Darwin and the creationists, also has an ancient origin. When did this age-old battle begin? When in history did thought diverge into two paths, one materialist, one idealist? In Chapter 2, "Heaven and Earth," we meet the first materialists, whose work divines the shape of things to come; the Atomists believed in evolution, and an atomic world of matter in motion, with no God.

What bearing does the ancient war between Athens and Sparta have on the development of science? The Peloponnesian War that reshaped the ancient Greek world is key to Chapter 3, "The Darkness Rising." The strong slave-owning city-state of Athens was brought under the iron fist of Sparta. Reduced to subjection, what effect did the siege have upon the idealist thinkers Plato and Aristotle, and how did a chronic fear of change develop into a philosophy of reaction? The chapter uncovers the divorce of thought from experiment and the curse it laid on science until the days of Galileo.

The eve of each revolution is played out in Part II, "The Gathering Storm."

In this section of the book, God's divine hand is adjusted to the great plan of a universe conceived by the Greeks, as the wisdom of the ancients is tailored to the medieval Holy Church. In following the trail, we consider how once fluid ideas were malformed into rigid ideology, how Aristotle's hypotheses were distorted into dogma, and how Plato's visions were weaved into theology.

In Chapter 4, "The Medieval Sky," how did the Aristotelian universe become crucial to the Christian drama of life and death? And why was it so criminal for Copernicus to move the Throne of God? In Chapter 5, "The Great Chain of Being," we consider the *scala naturæ*, the grand scheme of the universe, a strict hierarchy, in which every object and creature had its exact assigned place. Why did the notion of the missing link and of extinction become so apocryphal?

The drama of the revolutions is played out in Part III, "The Revolutions: The Weapons of Discovery."

Here is the part that Galileo and Darwin play in the unfolding drama. We examine the myths of their contrasting but starring roles. Galileo the first cynical scientist, Darwin the last gentleman philosopher; Galileo the abrasive experimenter, Darwin the placid naturalist; Galileo's search for meaning in the sky, Darwin's hunt for the secrets of life on Earth; Galileo's plea for an infinite space, Darwin's campaign for an ancient Earth. But both bound by fate, both defending incomplete theories, both tethered by their times.

The drama is played out on the largest imaginable stage. For our story is no naïve narrative of great discoverers, each with their own momentous and revelatory insight into the secrets of nature, dreamt up out of thin air. Such accounts of their work are too narrow and too idle to realize that "great men" are products of their time. Only by recognizing that they are men subject to the same sway of social influence, the same sorry compulsions, is their plight realised and their stature enhanced.

The year 2009 is a cause célèbre, a watershed for the weapons of discovery. The year marks both the 400th anniversary of Galileo's discoveries with the telescope, and the 150th anniversary of Darwin's theory of natural selection.

The celebrations afford an excellent opportunity to probe deeper into history, beyond those silly and conservative histories of science, rooted in the "great men" myth. It again allows a chance to ask questions about the conditions of culture and economy that led to such remarkable histories. Why Galileo, and why Darwin? What part did political bias and unconscious prejudice play in their creativity? And why did one famous scholar once say, "Darwin is truly great, but he is the dullest great man I can think of?"

So Part III examines in detail the different weapons of discovery that each man wielded in the name of science. In Chapter 6, "The Telescope and Galileo," we discover whether the road to modern physics really did begin in 1609. What was the relationship between the Galilean revolution and the war between the roundheads and cavaliers? And what was the link between Galileo and a revolution in the economy just beginning? In Chapter 7, "Evolution and Darwin," we find out whether the road to modern biology began in 1859. What was the link between Darwin and an economy just about to reach its highest expression? And, for that matter, why did physics come before biology?

The fallout of the revolutions is considered in Part IV, "The Aftermath: Worlds Turned Upside Down."

In Chapter 8, "The 'Galileo' Aftermath," we look at the impact of Galileo's discoveries on science, society, and state. If Galileo was not the great martyr of the new science, then who was? How did Galileo's ignorance of the latest findings of astronomical research lead to his eventual downfall? How did Galileo's connections with the nobility render his treatment by the Church more lenient? And how did his use of the telescope lead to the invention of science fiction and give impetus to the idea of alien life?

"The 'Darwin' Aftermath" is explored in Chapter 9. What has research uncovered of Darwin's role in the development of the theory of natural selection, in the 150 years since its inception? Did Darwin deceive his way into becoming the foremost evolutionary theorist at the time since becoming mistakenly identified throughout the world as arguably one of the most influential theorists in the history of science and civilization? And what role did collusion and class play in Darwin being associated with the theory?

Lastly, in Part V, "The Prestige," how do the two revolutions continue to shape the twenty-first century?

We live in a troubled time. A time when the rational skepticism of science is ridiculed by the certainty of faith. A time when the details of Darwinism are challenged by creationism, and when the unraveling human genome presents new problems for the morality of science. A time when big bang cosmology constantly seems on the point of collapse, and the further exploration of space is called into question. And a time when the independence and autonomy of scientists is challenged even further by the rise of corporate technoscience. How can the tale of Darwin and Galileo help in the future?

PART I

WISDOM OF THE ANCIENTS

CHAPTER 1

GREEK SKY: THE WORLD BEFORE THE TELESCOPE

THE CREATION

One of the most enigmatic works in the history of European art is to be found in Madrid's magnificent Museo del Prado. The Prado museum and gallery contains a world-class collection, including, naturally, the world's finest collection of Spanish painting—Velázquez, Goya, as well as El Greco, and most other leading Spanish old masters. But the gallery also holds the most important work of the medieval Dutch painter, Hieronymus Bosch.

Bosch's *The Garden of Earthly Delights* is a triptych, painted around 1504. The work is an orgiastic vision; its topic is biblical creationism and the consequences of sin. Its narrative arc is a journey in time, read from left to right. It begins with the creation of Eve, followed by earthly sin, and culminating in damnation. Through the use of vivid symbolism, the painting portrays a cornucopia of images, teeming with wriggling naked bodies, giant songbirds, hybrid beasts and bizarre objects, all of which seem to evade interpretation. For Bosch is the black hole of art. His works swallow any meaning you throw at them, their sheer surreal complexity leading to a wide range of interpretations from many scholars over the ages.

But *The Garden of Earthly Delights* has one of art's greatest flip sides. For, once the triptych wings of *Eden* and *Hell* are shut, *The Creation* is revealed. Now, the world is presented as a giant crystal globe, set against a black background.

The strangeness of the geography is striking. This world is both flat and round. A level and circular landmass, surrounded by a ring of ocean.

Earth and water fill the bottom hemisphere, like a dumpling sitting in stew. The airy top half is filled with dark rain clouds and the odd shaft of sunlight. The vision is an episode from the creation of the world, the dividing of land and waters. And it is God's world. There he sits, in the top left-hand corner; to one side a Latin inscription from Psalm 33 reads, "Ipse dixit et facta sunt; ipse mandavit et creata sunt" (He spoke, and it was done; He commanded, and it was made).

Like God, we look down from a superior height. This "world-egg" appears as a fully controlled environment, like a snow dome, or a terrarium. It is a constructed world, a plaything. In contrast to the multicolored vista of *The Garden*, with its shocking pinks and electric blues, *The Creation* is painted in grisaille, a gray monochrome. For this is a world fashioned and set in stone. God's word is gospel. And His world is contained and constrained inside a sphere, a world with hard limits in space and time.

In earlier centuries many thought Bosch's work was inspired by medieval heresies and shadowy hermetic arts. While the art of the older masters was based in the physical world of everyday experience, Bosch was master of the monstrous, discoverer of the unconscious. By the twentieth century, scholars came to view Bosch's visions as less fantastic, recognizing in his art the orthodox religion of the age. The depictions of writhing sinful humanity, the conceptions of Heaven and Hell.

So, Bosch may be the black hole of art, open to interpretation, but one thing is clear: *The Creation* is typical of the belief systems of its time. God's divine hand, adjusted to the great plan of a universe conceived by the Greeks.

OUR VIEW ON THE UNIVERSE

Before allowing Pythagoras to haul us into history, and to the origins of the great plan of the universe depicted by Bosch, it would be best to first sum up what is known today about the cosmos. A brief state-of-the-Universe report, if you like. Our method of interpreting the Universe around us is a rational one, originating with the Greeks. Modern cosmology, the study of the cosmos as a whole, is a science. And it is important to note that science provides the *best current interpretation of the natural world*. So it is worth taking a little time out first to examine just what "the best current interpretation of the natural world" means.

By "best," we imply that there are, of course, competing theories in science, competing explanations of the way the natural world is seen and understood. This was certainly true of Greek "science"; it provided more than one model of the ancient sky, as we shall see. By "current," we imply that science never stops growing. It is cumulative. Our understanding is not a static one, and that means our theories will also change in the future. Science is an ever-growing body of knowledge, built up from sequences of the reflections and ideas, the experience and actions, of a great stream of thinkers and workers. Science is permanently under repair, but always in use.[1]

Finally, by "interpretation," we mean exactly that. Science is not gospel. Scientists build models of hypothetical worlds and then test their theories. What is right about a model one day, may be wrong the next. In fact, some thinkers, most notably the Austrian philosopher Karl Popper, have suggested that a theory is scientific only if it is falsifiable!

These aspects distinguish science from the other great human institutions, such as religion, law, and art. Religion is concerned with the preservation of "eternal" truth. Science, on the other hand, is critical. The scientist deliberately strives to change accepted truth, by reference to provable and repeatable observations in the natural world.

A Warning on "Great Men"

This book, starting with Pythagoras and continuing on to Galileo and Darwin, will present an unconventional view of the advancement of science. The conventional view of the evolution of science is that of a peaceful, smooth, and uninterrupted development. It is both one-sided and false. In truth, the progress of science involves two essential elements. The first is long periods of investigations, which rely upon a gradual advance on tradition and custom. This first element is the fruit of many ordinary thinkers and workers. The second element is the revolutionary tipping points. In spite of all gradualness, an innovatory leap leads to decisive change. This element is usually associated with the "great men of science."

But there is a problem with this "great men" myth. It has led to a false idea of science, one which suggests that progress in science is due solely to the genius of great men, irrespective of factors such as culture, society, and economy. We are expected to believe that these masterminds just dream this stuff up out of thin air. Many conservative histories of science are rooted in the great men myth. They are little more than a series of naïve narratives of great discoverers, each with their own momentous and revelatory insight into the secrets of nature.

Let us be clear: great men such as Darwin and Galileo have been a crucial factor in the development of science. But their contribution should be studied in context, and not in isolation from their contemporary social setting. An inability to see this often leads to the use of redundant words like "brainwave" or "genius" to explain away those eureka moments of discovery.

In truth, conventional narratives devalue great men. Such accounts of their work are too narrow and too idle to realize that great men are products of their time. Only by recognizing that they are men subject to the same sway of social influence, the same sorry compulsions, is their plight realized and their stature enhanced. Indeed, as we shall see, the greater the man, the more he is immersed in the milieu of his days. He becomes more important. For only by seizing the moment chanced by his times is he able to make that innovatory leap that leads to critical change.

Nor is the great men myth specific to science. It applies to thinkers in any cultural field. But the hold of the myth on histories of science retained its grip far longer than with histories of culture and society. No effective discovery can be made in any field, be it biology or astronomy, without the necessary groundwork of thousands of relatively minor workers. It is on the basis of this painstaking work that great men make crucial discoveries.

Societies produce populations with a vast array of mental abilities. Relatively few seek a career in science, though more have the opportunity of doing so in the twenty-first century than ever before. Scientists are likely to differ greatly in their characteristics. But this much they will have in common: they are creatures of the culture in which they swim. Differing individual mental faculty gives a great variety to science. But imposed social and cultural controls and influences give science its unity, and makes possible the collective effort of science to understand and master our environment.

STATE-OF-THE-UNIVERSE REPORT

Bearing all this in mind, let us get back to our state-of-the-Universe report. We have found that there are more stars in the Universe than there are grains of sand on all of the beaches on planet Earth. It is worth pausing for a while to imagine being on one of those beaches, soft golden sand, sweeping off into the distance. Bright, sunny day, of course. In fact, let us make it the Caribbean. No expense spared. You reach down, hands cupped, and gather up two handfuls of the golden sand. Then you let the sand fall through your fingers, the grains glistening as they catch the sunlight. Each grain is a star. And each star is a sun, like our own local star, the Sun. You saunter on a few more steps, and again you gather up the sand and let it fall. And so on, over all the beaches on Earth. So much sand, so many stars.

On a large scale, swarms of such stars dwell in galaxies, effortlessly wheeling their way through the vastness of deep space. Each galaxy contains millions, if not billions, of stars. Under cover of the night sky, some galaxies can be seen with the naked eye. It is worth remembering that the very word "galaxy" derives from the Greek term *galaxias*, "milky circle," for its appearance to the eye.

Imagine yourself again, on our same trip, under the starry reach of the Caribbean sky. You look up, the star field is dazzling. But here and there is the odd nebulous smudge of a galaxy. The great galaxy in the constellation Andromeda, for instance, one of our near neighbors, and yet even this nearby citadel of stars is resolvable only by telescope. So that is how galaxies look to the naked eye, like tiny Tipp-Ex thumbprints in the sky. And yet the Milky Way Galaxy holds between 200 to 400 billion stars, and has special significance since it is the home galaxy of planet Earth. The capitalized word "Galaxy" refers to our own Milky Way, to distinguish it from the hundred billion or so other galaxies in the observable Universe.

Since the days of Galileo and before, it had long been suspected that stars other than the Sun are orbited by their own planetary systems. Indeed, extrasolar planets, planets beyond our Solar System family of eight planets, finally became a subject of proper scientific study in the mid-nineteenth century. But only fifteen years ago did the tipping point occur. For the first time, confirmed findings of extrasolar planets were made. To the naked eye, the extrasolar planets themselves cannot be seen. But their parent stars are clearly visible. Indeed, our Caribbean sky would once more reveal its secrets. Stars in the constellations of Ursa Major, Andromeda, and Pegasus, for instance, each have stars around which extrasolar planets are known to orbit.

And it is not just the Caribbean sky. From any garden on Earth, where these constellations are visible, these extrasolar planets orbit, unknown to the naked eye. To date,[2] 344 extrasolar planets have been found in the local solar neighborhood of space. Many believe that around one in ten of all Sun-like stars have planets. The true proportion may be far higher and, if that is true, the universe may be replete with planets, heightening the possibility that some might support extraterrestrial life.

And all of this, Solar System and Sun, countless stars and extrasolar planets, Galaxy and myriad galaxies, is set adrift in an expanding Universe so immense that light from its outer limits takes longer than twice the age of the Earth to reach terrestrial telescopes. We live in a changing Universe, one that has been expanding since the famed "big bang," the dawn of time. A Universe that continues to expand, into the infinite reaches of space.

Lastly, to help us get a grasp on the sheer scale of the Universe, consider time travel. For stargazing is a kind of time travel. Since light, the fastest thing known to science, takes time to make any journey from A to B, it makes sense that light also takes time to cover the vast distances involved in a journey through space. For example, the nearest star to Earth, Proxima Centauri, is about 25 million million, or 25 trillion, miles away. Light takes more than four years, from that nearby star, to hit the naked eye. Modern astronomers say that Proxima Centauri is four "light years" away, since that is how long it takes light to make the journey. We are looking at a four-year-old image of the star.

At any moment, we see the sky as it was in the past. The further we look out into space, the further we look back in time. So, it follows that we can use light to scale space, to get a measure of the Universe. A light journey from the Moon to the Earth takes light around one second. We see the Moon as it looked a second ago. The Moon is a "light-second" away, we say. Similarly, a light journey from the edge of our Solar System to the Earth takes around one light-year. We are around 28,000 light-years from the center of our Galaxy, and one of our nearest galactic neighbors, the Andromeda galaxy mentioned earlier, is about two and a half million light-years away. We see Andromeda as it appeared two and a half million years ago, because its take light that long to make its journey to Earth.

As to what space actually looks like, in terms of its large-scale structure, we cannot really be sure. Astronomers are trying to characterise the distributions of matter and light on the largest scales, typically on the order of billions of light-years. The most accurate sky surveys to date suggest that clusters and superclusters of galaxies seem to be arrayed into gigantic cosmic domains that resemble the cells and filaments of a sponge. Larger than this, there seems to be no continued structure, a phenomenon which has been called "the End of Greatness."

It's little wonder Douglas Adams started his famous *Hitchhiker's Guide to the Galaxy* by jauntily telling his readers they, "won't believe how vastly, hugely, mind-bogglingly big" space is. All of this was unknown to Galileo, when he first gazed up at the sky through the telescope. But we'll leave our pencil portrait of the universe for now, our brief state-of-the-Universe report. For first we must delve into Galileo's backstory, and the history of the ancient sky.

PYTHAGORAS: THE MOVIE

Pythagoras hauls us into a history of astronomy. Listen, forget about his dreaded theorem. Understandably, many will have been tormented by his triangle during their schooldays, but there is so much more to the man, and more importantly, the influence of the Pythagorean Brotherhood. The very word "philosophy" is Pythagorean in origin. When we use the word "harmony" in its wider sense, when we call numbers "figures," we speak the tongue of the brotherhood. And their approach was epoch-making; through their application of mathematics to the human experience, they were founders of what the world understands today as science.

The Pythagorean Brotherhood was founded in the seminal century of awakening, the sixth century BC. Elsewhere on the globe, Siddhārtha Gautama began the spiritual teachings in ancient India that led to the foundation of Buddhism. And the teachings of the Chinese thinkers Confucius and Lao-tze had begun to deeply influence Eastern life and thought. Pythagorean astronomy was part of an all-embracing philosophy. The brotherhood tried to synthesize an interconnected view of the universe that incorporated religion with science, medicine with cosmology, mathematics with music—mind, body, and spirit as one.

Pythagoras himself was born sometime between 580 and 572 BC on the Greek island of Samos. Set in the North Aegean sea, Samos is close to Miletus, an ancient city on the western coast of what is now Turkey. Miletus was a seat of considerable learning that produced notable ancient philosophers such as Thales and the atheist Anaximander, the teacher of Pythagoras. There is some dispute as to whether Pythagoras truly was an entirely legendary figure. But that need not concern us here. The brotherhood that bore his name was to have a massive influence on Greek cosmology and thought, particularly through its most prominent exponents, Plato and Aristotle.

Since very little is known about Pythagoras, as none of his writings survive, it is impossible to say whether a specific detail of the Pythagorean worldview was the work of Pythagoras himself, or that of one of the brotherhood. Many of the teachings associated with Pythagoras may actually have been those of his peers or successors. And that is just how it should be. For many regard Pythagoreanism as the first expression of collective, democratic thought. Women were given equal opportunity to study as Pythagoreans, though they were also encouraged to learn practical domestic skills in addition to philosophy, being held to be different from men, in good ways as well as bad.

Pythagoras seems to have been the first to call himself a philosopher, a lover of wisdom. Around 530 BC, he migrated south from Samos to Croton, in southern Italy, and founded a secret philosophic and religious school. His reputation must have preceded him. For soon after his arrival, the brotherhood ruled the town. They went on to dominate a major part of Magna Graecia, Greater Greece, the area in southern Italy and Sicily that was colonised by Greek settlers from the eighth century BC, and who brought with them an enduring stamp of their civilization. But, in the case of the Pythagoreans, their secular power was fleeting. They were exiled from Croton, their temples razed to the ground, members of the brotherhood butchered.

Two tendencies merge in Pythagorean philosophy, the *mathematical* and the *mystical*.[3] Many are rightly skeptical of whether Pythagorean mathematics was truly original. His infamous theorem on the right-angled triangle, for instance, was known to have been a practical rule that served the Egyptians and Babylonians well before the brotherhood came along. There is even evidence to suggest that the entire number theory of the Pythagoreans, in both its mystical and mathematical facets, was derived from Eastern thought. But synthesis was paramount. The brotherhood produced a fusion of mathematics, science, and philosophy that has had a lasting influence on the human race.

PYTHAGOREAN ASTRONOMY

The Pythagoreans recognized that numbers were the key to comprehending the cosmos. It is important to realize that the brotherhood were not, in their view, reducing the human experience. Early in the nineteenth century, Romantic poet John Keats famously explored the tension between reason and sensation in his poem, *Lamia* (1819):

> Do not all charms fly
> At the mere touch of cold philosophy?
> There was an awful rainbow once in heaven:
> We know her woof, her texture; she is given
> In the dull catalogue of common things.
> Philosophy will clip an Angel's wings,
> Conquer all mysteries by rule and line,
> Empty the haunted air, and gnomed mine—
> Unweave a rainbow

But to the Pythagoreans, philosophy began in wonder. Philosophy was the highest music, and the highest form of philosophy was concerned with numbers, for ultimately all things are numbers. So, finally, when philosophic thought had done its best, the wonder would remain. So rather than mathematization leading to a reduction of human experience, it was an enrichment.

The Pythagorean concept of "harmony" was typical of the way in which the brotherhood synthesized an interconnected view of the universe. Numbers were not tossed into the world at random. They were arranged, or arranged themselves, like the structure of crystals, like a musical scale, according to the universal laws of harmony.

The basic Pythagorean notion of *armonia* (harmony), regarded the human frame and body too as a kind of musical instrument. Each string within must have the right tension, the correct balance, for the patient's soul to be in tune. The musical metaphors that are still applied to medicine, such as "tone" and "tonic" and "well-tempered," are also part of our Pythagorean heritage.

HARMONY OF THE SPHERES

The brotherhood's focus on harmony was extended from the body and soul of man, on one side, to the stars, on the other. The ancient Egyptians, Babylonians, and Hebrews had considered the universe to be a kind of cosmic oyster. Roughly speaking, each ancient cosmology imagined a heavenly arrangement, with water underneath, water overhead, supported by a solid firmament. What the Babylonians and Egyptians began, the Greeks refined. But their refinement was rational, rather than mythological.

Pythagoras's teacher, Anaximander, one of the foremost philosophers of Ionian Greece, had developed one of the first accounts of a mechanical model of the universe. His cosmos was no closed oyster. It was a universe infinite in space and time. The matter that made up this universe was no ordinary material. It was indestructible. It was eternal. And out of this stuff, all things were made, and into it they returned, like a cosmic recycling scheme.

At the center of this cosmos of Anaximander's, the Earth was set adrift. No longer the ancient idea of an earth-disc floating in water, as we saw in the medieval portrait by Bosch. Anaximander rather surreally considered the Earth to be a cylindrical column. Bizarre, perhaps, but his truly bold innovation was to surround this cylindrical Earth with air, floating upright at the center of the universe, without support or structure. Yet the Earth did not fall. Being at the center of all things, it had no preferred direction in which to fall, for if it did, it would disturb the symmetry and balance of the whole.

In the Pythagorean universe, the cylindrical Earth of Anaximander becomes a sphere. Around this central Earth, and moving in concentric circles, revolve the Sun, Moon, and planets. Each heavenly body was fastened to a sphere, and the fleet revolution of each of these bodies caused a musical stirring of the air. In the view of the brotherhood, each planet would hum with its own unique pitch, for its sound would depend on the ratio of its particular orbit around the Earth. Like a huge and heavenly musical instrument, the orbiting planets too were subject to the universal laws of harmony.

In his famous *Natural History* (circa 77 AD), the ancient Roman scientist and nobleman Pliny the Elder reported that the Pythagoreans considered the musical interval formed by the Earth and Moon to be a tone; Moon to Mercury, a semitone; Mercury to Venus, a semitone; Venus to the Sun, a minor third; Sun to Mars, a tone; Mars to Jupiter, a semitone; Jupiter to Saturn, a semitone; and Saturn to the sphere of the fixed stars, a minor third. The "Pythagorean Scale" so produced from this musical interval is still extant. But for us, Pliny's report reveals not only a heavenly musical scale, but also a cosmic architecture that was to have a profound influence on the history of astronomy.

PYTHAGOREAN SKY

Legend had it that only the master of the brotherhood, Pythagoras, possessed the gift of actually hearing this music of the spheres. But the vital development here, which led later to the crowning achievement of Pythagorean cosmology, was the implied structure of the universe. Thousands of years before, the Babylonians had been avid sky watchers who scanned the stars and mapped the heavens. Clay tablets from the reign of Sargon of Akkad, of around 3800 BC, illustrate an already well-established astronomy.[4] Precision was crucial. They had measured the length of the year to within 0.001 percent of the correct value.[5] Though their cosmology was based on mythological notions, their calculations enabled them to prefigure the seasons and the rains, harvest and sowing time, the regular cycles of the year and their associated religious ceremonies. Essentially, their theory worked.

From these ancient observations, the Pythagoreans inherited a firmament. The stars remained stationary, like pinholes in a dark fabric through which was glimpsed a cosmic fire beyond. Like the Babylonians, the brotherhood may have derived a feeling of safety and security from the utter dependability of the heavens. Immutable and predictable through the ages, the stars in their fixed patterns and imagined constellations must have contrasted greatly with the turbulent lives of the sky watchers below.

The sphere of fixed stars was not the only fascination inherited by the Pythagoreans. A number of so-called vagabond stars had in equal measure beguiled and perplexed sky watchers for many generations. To the ancient eye, without the use of a spyglass, only seven of these "wanderers," or "planets" as they were known, could be seen among the thousands of lights that bejewelled the firmament. The wanderers were different. True, like the fixed stars, the Sun, Moon, Mercury,

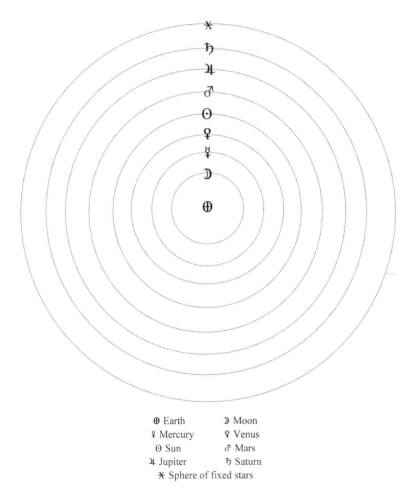

⊕ Earth	☽ Moon
☿ Mercury	♀ Venus
☉ Sun	♂ Mars
♃ Jupiter	♄ Saturn
✳ Sphere of fixed stars	

Figure 1.1 The classical geocentric system of the Pythagoreans. The planets orbit counter-clockwise, as viewed from above.

Venus, Mars, Jupiter, and Saturn, all seemed to revolve once a day around the Earth. But the planets also had a peculiar motion.

These seven bodies wandered and strayed along the path they described across the night sky. Yet they did not rove about the entire heavens. Their esoteric behavior was limited to a narrow strip of sky, a belt that encircled the spinning globe of the Earth at an angle of about 23 degrees to the equator. This belt, the Zodiac, was separated by the ancients into a dozen parts, and each part named for the constellation of fixed stars in that region of the Zodiac. And along this belt the planets roamed.

For the mythologically inclined sky watchers, the passing of a planet through a constellation of the Zodiac had twin connotations. It provided figures for their meticulous observation and supplied symbolic messages of ritual significance. For the more rational Pythagoreans, with their rule of harmony and number, a structure of the universe emerged.

In the Pythagorean system, the planets moved with the same regularity as that of the rotating sphere of fixed stars beyond—in circular orbits around the central Earth. This gave a good account of the observations of the behavior of the Sun on its yearly journey through the plane of the ecliptic, the apparent path of the Sun across the heavens. Their system also gave a reasonably accurate account of the rather less regular motion of the Moon. But the circular orbits got nowhere near explaining the observed motions of the other five wandering planets. Of which, more will be discussed later.

Nonetheless, once the position and shape of the orbits was established, it was possible to make an educated guess on the layout of this ancient solar system. Planets such as Jupiter and Saturn described a slow motion across the sky, appearing to keep up with the fixed stars beyond. Today, we know that it takes Jupiter roughly twelve years and Saturn around thirty years to make a complete journey around the central Sun. To the ancient eye, since these planets very nearly kept pace with the stars, they were assumed to be far from Earth, and close to the stellar sphere that bounds this universe.

In contrast, the Moon loses 12 degrees each day in its apparent race with the stars. The Pythagoreans must have seen this justified the suggestion that the Moon was closer to the Earth, which was assumed to be stationary, as well as at the center of the system. So, the outer limit of this cosmos was the stellar sphere, and just inside was Saturn, since it was the planet that took longest to move around the Zodiac. Next came Jupiter and Mars, arranged in order of decreasing orbital period, the time taken to make one complete orbit. Innermost was the Moon, since the lunar orbit placed it closest to Earth.

But the remaining three planets, the Sun, Venus and Mercury, posed a problem. All three vagabond stars made their apparent journey about the Earth in the same common time, one year. Their order could not be unpicked with the technique used for the other planets. Indeed, there was much disagreement on this matter among the philosophers. The layout shown in figure 1.1—Sun outside, followed by Venus and Mercury within—was not undisputed, but for reasons now lost in the mists of time, was the most popular order.

Indeed, the layout of the system held a remarkable potency and power, which was to last until the days of Galileo.

Consider the science communication of this system. Figure 1.1 illustrates almost all the knowledge enjoyed by the non-astronomer in ancient times. The term "inferior planet" was now used for those planets (Mercury and Venus) that were between the stationary Earth and the orbiting Sun, and the phrase "superior planet" was used for those planets (Mars, Jupiter, and Saturn) that lay beyond the Sun's orbit. The diagram gives no indication of the dimensions of the orbits, no account of the irregularities of the planets in motion. But these further developments of ancient astronomy, as we shall see, were too mathematical for most laymen to grasp.

Now consider astronomical research. The layout proved a potent tool, at once efficient and fertile. For one thing, the notions embodied in the diagram gave a satisfying account of both the phases of the Moon and lunar eclipses. The Pythagoreans had used these eclipses as part of their reason for believing that our world was spherical; the shadow cast by the Earth upon the Moon during these events was, after all, round. And the only solid that always projects a round shadow is a sphere.

Later, the philosophers that followed the brotherhood used the concepts of this model to make measurements of the universe that were quite startling in their accuracy. For example, Eratosthenes measured the Earth's circumference with unerring precision during the third century BC. And during the second century BC, Aristarchus, of whom much more will come later, brilliantly calculated the sizes and distances of the Sun and Moon!

These examples give a brief taster of the power and ingenuity of ancient Greek astronomy, inspired by the Pythagoreans. We shall discuss this further in due course, but for now the potency of this simple model of the universe should be duly noted. Not only did the model make common sense to the simple naked-eye observations that were made nightly, under a Greek sky that seemed to revolve around the central Earth. The system could also be used to predict important events such as eclipses, those ominous cosmic encounters that had previously given the ancient mind such pause for superstitious thought. It is hardly surprising that the model became a rational tool, a conceptual scheme with an increasing hold upon the minds of both astronomers and laymen.

THE BIRTH OF THE RATIONAL

Now would seem like a good time to take the opportunity to address some important questions, the kind of questions that never seem to get asked in conventional histories of science. Nobody before the Pythagoreans had considered that mathematical associations might hold the mysteries of the cosmos. And in general it is the Greeks we have to thank for the conscious and unbroken thread of history and science that begins with the dawn of rational thinking in the ancient classical world. So, why *then*? Why in ancient Greece does all this develop? And what were the conditions of culture and economy that led to such a remarkable history?

It began with the massive progress made with the foundation of Greek cities. The features of science, which distinguish it from other aspects of human social activity such as those of art or religion, are concerned mostly with the tradition and technique of how to do things. Each work of science functions as a recipe: it tells you how to carry out certain things if you want to do them. Science is not just a matter of thought. It is the process of thought continually carried into action, continually refined by practice. This is what gives science its crucial link to the means of providing for human needs, and it is why science should not be studied separately from technique.

In the history of science, we repeatedly see new aspects arising out of technique and new developments in science giving rise to new branches of technique. According to legend, Pythagoras discovered the link between music and maths from a blacksmith. One day, he was passing a smith at work. Hearing the sweet sound of the smith striking the anvil, Pythagoras realized that such harmony must bear some relation to mathematics. He spent some time with the smith, examining the tools and exploring the simple ratios between tools and tones.

With cities came the challenge of providing for human needs on a massive scale, and with that came the necessity of innovation, in science and technique. Once this provision had been met, next came the challenge of the associated large-scale administration.

In the ancient cities, the responsibility for this considerable administration lay with the priests. They had exclusive access to the means of measuring, calculating, and recording. Indeed, it is worth remembering that the term hieroglyphics means "priest's writing." It is a reminder that science has, from ancient times, been associated with a governing elite. In the minds of the great mass of people, a restricted access to science engendered a deep mistrust of science, and book learning in general. Such was the case five thousand years ago with the first cities.

But the Greek cities were radically different. For one, they were far more democratic. For in their lands, between the twelfth and sixth centuries BC, the Greeks forged a unique culture. Theirs was the most fruitful exploitation of the emergent Iron Age. They were relatively isolated from the conservative sway of the older civilizations, protected from invasion by their initial poverty, and girded by their burgeoning sea power. Meanwhile, they were still able to learn from the traditions of the more ancient civilizations. They took from these foreign cultures many practical techniques and various accounts of the workings of the cosmos.

The Greeks did not as much invent civilization, nor even inherit it. They discovered it. They were alone in appropriating the vast bulk of learning that was still extant, after centuries of war and abandon in the ancient empires of Egypt and Babylonia. What was obscure, the Greeks ignored; what was superstitious, they disregarded. They procured the ancient knowledge and grasped an opportunity. With keen intelligence, they forged a new culture, at once simpler, more abstract, and more rational.[6]

The resulting classical civilization that emerged has remained one of the keystones of today's world culture. Its greatest contributions were in political democracy and natural science, particularly mathematics and astronomy. Indeed, the Greeks were struck with the esteem of the noble sciences of the older ancient civilizations; it influenced them significantly. The subjects of mathematics, astronomy, and medicine remained the foundation of education throughout the classical Greek period and beyond, into the medieval period. The ignoble sciences, such as chemistry and biology, have still to struggle for the same cultural respect.

And now, we can begin to understand why these conditions of culture and economy hastened the dawn of rational thinking. Take, for example, the archetypal Greek city state of Attica, containing Athens. The region was not rich in corn growing. It was dependent on exports of olive oil, silver, and pottery to feed Athens' relatively immense population of three hundred thousand. But the compact city was able to efficiently marshal its resources to the full. In an environment such as this, there were rapid, if not violent, changes in economy and politics. Tradition was at a discount. Those citizens with the motivation and nous were rewarded with an improved standing in society, without the fetters of clan or tribe. Institutions and divinities became less significant; men were more important than gods.

The *dialectic* was born. The lifting of ancient sanctions meant that cases had to be argued out on their merits. And the history of Greek science, or philosophy as it was in those days, is the history of a series of back-and-forth arguments, called the dialectic. The other crucial factor is this: the very ability to make such arguments was extensively promoted by the intense political culture of everyday Greek life. The importance placed on deals in trade and law, where each man represented himself and judges were chosen at random, led to the required development of debate to the highest possible level. As a result, the Greek compact city-states allowed far superior prospects for the typical citizen than did the capital of a great empire.

So began the rational philosophy of the Greeks. In this context we can now view the Pythagoreans in another light. The brotherhood was part of the first phase of Greek science, concentrated in Ionia, the region that felt the influence of the older civilizations most keenly. We can further see the Pythagorean school as one of the first expressions of democratic thought. Their philosophy, fitting for an age of progress, was positive and optimistic. It was also materialist, reasoning on what the world was made of, and how it had come to be. And their rationalism can be seen as part of a rising mercantile class, who were opposed to an orthodoxy of empire and landed nobility.

THE BROTHERHOOD BITES THE DUST

The original Pythagorean community was political in the way we have described above. And as such they were persecuted and dispersed. Some aspects of their order were that of a religious school. They followed an ascetic life, sharing all property, and leading a communal existence. Like primitive Christian communities, they led a life that obeyed ritual and restraint. They devoted time and energy to meditation and the contemplation of conscience. The Pythagoreans were also a force in Italian politics. According to reports, after their first sermon to the Crotonians of southern Italy, six hundred joined their communal life, without even returning home to bid farewell to their families.[7]

And yet the brotherhood was a very influential academy of science. They understood that reality could be reasoned into number. Indeed there is evidence that Pythagoras's countrymen were also well aware of the great technical potential of geometry. The Greek historian Herodotus, regarded in the West at least as one of the foremost of ancient historians, reported in the fifth century BC that the islanders of Samos were,

> the makers of the three greatest works to be seen in any great land. First of these is the double-mouthed tunnel they pierced for a hundred and fifty fathoms through the base of a hill … through which water, coming from an abundant spring, is carried by its pipes to the city of Samos.[8]

For two millennia it seems this account was taken with a pinch of salt until, at the beginning of the twentieth century, the tunnel was actually discovered and excavated.

Running at a full 823 meters long, complete with watercourse and inspection path, the structure is an amazing feat of ancient engineering. More impressively still, evidence clearly shows the tunnel was begun from both ends, the first in history to be constructed in such a methodical way. The watery subway was built under the knowing supervision of Eupalinos, one of the first hydraulic engineers in history whose name has been passed down. Indeed, only recently a huge road tunnel, named after Eupalinos, has been driven through the Geraneia mountains in Corinthia to accommodate a new expressway link between Corinth and Athens. Such was the setting in which the brotherhood was born.

But the Pythagorean school was also greatly concerned with the morality of society. Pythagoreans were expected to live ethically, love one another, discuss politics, practice pacifism, and dedicatedly study the mathematics of nature. Given the example of Eupalinos's engineering feat, perhaps it is little wonder the brotherhood were well aware of the dialectic of science, its potential for man's deliverance and devastation. The Pythagorean diet was vegetarian, since only members pure in body and spirit should be trusted with nature's secrets.

And tragically, one of nature's secrets helped bring about the end of the brotherhood. For the Pythagoreans discovered "irrational" numbers. These numbers, such as $\sqrt{2}$ or π, are numbers that cannot be written down as the ratio of two integers, two whole numbers. For philosophers who contemplated that all of nature could be understood by number series and number ratios, this was a major blow. If philosophy was the highest music, and number the highest philosophy, we too can perhaps begin to grasp the problem.

The proof of the existence of irrational numbers is attributed to a member of the brotherhood, namely Hippasus of Metapontum. He is thought to have discovered them while analyzing the geometry of the pentagram, used by the Pythagoreans as a symbol of recognition among members and as a mark of inner health. At first other members tried to disprove the existence of such numbers through logic. They failed. After all, today we know almost all real numbers are irrational. Believing in the absoluteness of numbers, the Pythagoreans kept the discovery a secret, dubbing the irrational numbers "*arrhetos*," unspeakable. But Hippasus let the scandal leak, and legend has it he was put to death by drowning.

> It is told that those who first brought out the irrationals from concealment into the open perished in a shipwreck, to a man. For the unutterable and the formless must needs be concealed. And those who uncovered and touched this image of life were instantly destroyed and shall remain forever exposed to the play of the eternal waves.[9]

Dramatic indeed.

To make matters even trickier, the brotherhood disbanded. The exact reasons are unknown. But it seems the local authorities in southern Italy showed a growing intolerance to the egalitarian, if not communistic, philosophy and practice of the Pythagoreans. As an order they were persecuted and finally dispersed, though prominent followers who had earlier disappeared into exile, such as Philolaus, were later allowed to return to Italy and teach.

THE INFLUENCE OF THE BROTHERHOOD

The philosophical work of the Pythagorean Brotherhood is no less than the beginning of mathematics and physics, at least in the West. As one scholar said, "Pythagoras is the founder of European culture in the Mediterranean sphere."[10] Admittedly their mathematics was mystical, a form of idealism that lives with us still in the blessed trinity, the four evangelists, the seven deadly sins, and the number of the beast.[11] On the other hand, the mystical element of the Pythagorean "harmony of the spheres" seems to have had curious and perhaps capricious consequences.

It inspired Galileo's father, the lute player and maker, Vincenzo Galilei. Vincenzo discovered a new mathematical relationship between string tension and pitch. We can only speculate as to whether Vincenzo's work on the idea that music can be mathematically analyzed may have convinced Galileo himself to take an interest in physics. And Pythagorean harmony was an integral aspect of the work of the sixteenth century German astronomer and mathematician, Johannes

Kepler. As will be seen in due course, Kepler's mercurial talents helped lead to a modern understanding of the heavens.

At root, the Pythagorean harmony of the spheres was merely a manifestation of their mathematical approach. Like other thinkers and workers in their wake, they sought the secrets of the universe. To the brotherhood, as we shall increasingly see, we owe the very foundations of astronomy and cosmology. They recognized the importance of the circle and the sphere. They understood the Earth to be a sphere and realized that the planets moved in circular orbits. The next chapters shall reveal that Pythagorean ideas, when rationalized further by Heraclides and Aristarchus, led to an ancient picture of the cosmos very similar to our modern notion of the solar system.

The philosophy of the brotherhood stands at a crossroads. From it flows a legacy of two very divergent systems of thought. One pathway, that of Pythagorean number theory, was to be given a materialist twist by the Atomists, and lead eventually to modern science. This road of practical philosophy gave a means by which the physical world could be reduced to measure and number. Twenty-six centuries later, Western civilization is both blessed and cursed by man's resulting power over nature.

The other pathway, following the more abstract and mystic aspects of the Pythagoreans, laid the foundations for Plato's idealism. In physics, the Pythagoreans often ran too far beyond their actual experience. They substituted number mysticism for knowledge based on experiment. The brotherhood founded the mathematical process of proof by deductive reasoning. It is a powerful tool in maths. But in science, deductive proof can be used to prove conspicuous claptrap from allegedly obvious principles.

In the next two chapters we shall follow both pathways. Along one we encounter ancient philosophers, such as Democritus and Epicurus, and their Atomist creed. We shall see, in the Atomists' almost Darwinian worldview, that there is no God, that life exists elsewhere in the universe, that nature evolves, and that in reality nothing exists but atoms and the void. Along the other pathway the contrast is stark. Here we meet the rival philosophical school associated with those equally great and creative thinkers, Plato and Aristotle. We will see that there is sociology as well as logic to science. Some bad scientific ideas can spread widely, at least for a while. And some good ideas lie dormant for years before finally catching on and colonizing scientific imaginations. Along this second route, Plato proceeds to divorce philosophy from observation and lays a curse upon astronomy that is not lifted for two millennia. Until Galileo and the telescope.

Philosophy and science in the ancient Greek world

Date	Technical developments	Political developments	Philosophy and science materialist tradition	idealist tradition
600 BC		Age of tyrants	Influence of ancient learning	
	Acquisition of Eastern techniques		*Thales* and the Ionian philosophers Materialist theory of the universe	
		Persian conquest of Ionia	*Pythagoras*, maths, and physical law *Heraclitus*, philosophy of change	
		Liberation from Persians		
500 BC	Mining and metal working		*Philolaus*, spherical Earth in orbit	
	Shipbuilding	Pericles in Athens	*Empedocles*, four elements	*Parmenides*, change illusory
	Architecture and sculpture	Peloponnesian War	*Democritus*, atomic theory	
		Athenian democracy		
400 BC				*Socrates*, the dialectic method
	City building on grid plan	Defeat and reaction in Athens		*Plato*, Idealism
			Heraclides, non geocentric cosmos	*Eudoxus*, heavenly spheres
		Triumph of Macedon		*Aristotle*, descriptive biology
		Alexander's conquests		
300 BC		Museum of Alexandria	*Epicurus*, atomic philosophy	

Date	Technical developments	Political developments	Philosophy and science materialist tradition	idealist tradition
	Geography data on Persia and beyond	Hellenistic influence spreads abroad		*Euclid*, geometry
	Development of engineering	Wars with Carthage	*Aristarchus*, rotating Earth, heliocentrism	
200 BC	Great spread of slavery	Roman control of the Greek world		*Hipparchus*, epicyclic cosmos
100 BC		Roman civil wars		
		Conquest of Gaul	*Lucretius*, atomic materialism	
0 BC		Jewish revolt	*Hero*, mechanics, steam engine	
		Spread of Christianity		
100 AD		*Marcus Aurelius*, philosopher emperor		*Ptolemy*, descriptive astronomy
	Water mills			
200 AD	Decline of city economy and trade	Crises and barbarian invasion		
300 AD		Condemnation of Arianism		
400 AD		Breakdown of Western Empire		

Chapter 2

Heaven and Earth

At the Brera Palace

Giovanni Schiaparelli was the man at the center of the Mars scandal. The problem began in 1877. Schiaparelli was an Italian astronomer whose specialism was the solar system. He had naturally observed Mars on a number of occasions previously and had described seeing what he called "seas" and "continents." But the trouble started one night when at the telescope Schiaparelli noted long linear features on the Martian surface, which he called "canali," meaning "channels."

As has now become legend, Percival Lowell, a willing Boston entrepreneur, became convinced that Schiaparelli had identified artefacts of a dark alien race. In several publications early in the new century, Lowell described the canals as clear evidence of a sophisticated civilization. They were using the canals, he insisted, to transport water from the Martian polar caps to the parched equatorial regions.

Schiaparelli had formerly studied at the University of Turin. But for forty years he did the night shift as astronomer in residence at the Brera Observatory in Milan. The observatory is part of the Brera Palace, built on the site of the fourteenth century Monastery of the Humiliated Monks. Sitting among the dark narrow streets of one of the most fashionable districts in Milan, the Brera is a complex of beautiful baroque buildings, which surround the Palazzo Brera, a grassy opening in the city backstreets. The jewel in the crown of the Brera Palace is the famous Pinacoteca di Brera, one of Italy's outstanding collections of Italian paintings, an outgrowth of the cultural program of the Brera Academy, another integral part of the Brera Palace.

As Schiaparelli sits patiently at the eyepiece gazing up at Mars, we ghost, down through the dark corridors of the Pinacoteca di Brera. We steal into the Casa Panigarola, a room on the first floor. Amid the masterpieces of the casa is a dark portrait of two men. One man weeps, as he wrings his hands over the world. The other man laughs. Both, it seems, are scholars, as the table before them is littered with books. Together they symbolize the wisdom of the ancients.

Many centuries before the Mars misfortune, one of these men had seen a world ruled only by matter in motion, a cosmos that evolved in time. The other, inspired by early notions of evolution, had imagined nothing but atoms and void, and, more than two millennia before Schiaparelli's imaginary canals, had taken life into deepest infinite space.

The Two Traditions

So the Earth was set adrift.

From the Pythagoreans onward, the idea took hold that the Earth was a sphere, floating freely in space. Now *The Histories* of Herodotus is thought to be the first work of history in Western literature. Written in the fifth century BC in the Ionic dialect of classical Greek, *The Histories* tells many tales of the Greek city-states of the time. Herodotus traveled widely around the ancient world, collecting stories for his book and reporting only what he was told. Would that all historians were as honest.

One tale in particular is very striking. Herodotus reports a rumor that far up in the north lived a people who slumber six months of each year. The account shows that the implications of a spherical Earth had already hit home. Early anecdotes of a "land of the midnight sun" speak strongly of an understanding of the polar night, the length of the Midsummer Day increasing from twelve to twenty-four hours as you go from the equator to the polar circle.

So let us follow the two pathways that lead us from our point of departure with the brotherhood. The fascinating history of Greek science is, of course, a continuous evolution. But for clarity of purpose, the story shall first focus on the materialist tradition. This chapter is titled "Heaven and Earth" for good reason: the all-embracing worldviews of the early Greek materialists include thoughts on both astronomy and evolution, predating both Galileo and Darwin by many centuries. Indeed, the cosmology of the Atomists was perhaps the first to propose that the universe was replete with extraterrestrial life, too.

On this materialist road we shall encounter the worldviews such as those of the pre-Socratic philosopher of change, Heraclitus (535–475 BC). His idea that "everything flows" is then adopted by the likes of Empedocles (490–430 BC), who conjures up a rather peculiar forerunner of Darwinian evolution. Along this way, we will also consider the Atomist cosmology of workers such as Democritus (460–370 BC) and Epicurus (341–270 BC).

In the next chapter we shall follow the other route. That associated with the tradition of the more abstract and idealist philosophers, including Plato (428–348 BC), the pupil of Socrates, Plato's student Aristotle (384–322 BC), and Ptolemy (83–161 AD).

The ancient and noble science of astronomy will be our guide and our measure. We shall use developments in astronomy as a benchmark against which we shall gauge the progress of Greek science, society, and culture. In the next two chapters, we will find that astronomy not only perfectly illustrates the contrast between the traditions, but that it proved to be a grindstone against which other leading edges of science, such as thoughts on nature and evolution, were sharpened.

THE CRUCIBLE

First things first: the Pythagorean backstory. As has already been seen, around the sixth century BC the pioneering Pythagoreans were part of an early stage in Greek science that had begun in the Ionian cities of Asia Minor, particularly Miletus. Here, contact with the beguiling ancient civilizations was closest. And the word spread quickly to the new Greek colonies in southern Italy, where the brotherhood was based.

The change to an Iron Age economy was critical. The stranglehold of the landed aristocracy was losing its grip. Power was being usurped by a band of local bosses, with the aid of the mercantile classes. It was a time of violent expansion, with colonies emerging all over the Mediterranean. The fact that the epicenter of this commercial change was at first the Ionian settlements of the eastern Aegean explains why they were at the forefront of the new philosophy.

It was a crucible of change.

And that change inspired new solutions to traditional questions. Now, instead of relying on ritual and superstition to fuel a worldview, the emergent Greeks produced a new paradigm. The picture of the world that developed was both simple and material, based on everyday life and labor. The people who considered such matters were known as "sophists," wise men. Only later did they become "philosophers," lovers of wisdom. They retailed the ancient knowledge, for a new market.

These sages, like Pythagoras, often established religious orders that were also philosophic schools. The most successful leaders became political advisers to a democratic chief or tyrant, in those days the word tyrant carried no ethical censure. They gave rational advice on every kind of topic. Indeed, it bestowed kudos on a regime to have a famous sage in tow. For example, the legendary Greek statesman Pericles (495–429 BC) was so served by the pre-Socratic sophist Anaxagoras (500–428 BC). Ironically, Anaxagoras was not wise enough. He was so fond of flouting popular convictions that he was forced to retire. Clearly, philosophy can be a dangerous game.

Little of these early sophists is known. Most of that which has been handed down has come from the oral tradition. In some cases a few fragments of the more prominent sages are gleaned from the works of Plato and Aristotle, who used them mostly as either fools or foils. Plato, for example, derided the philosopher Protagoras (490–420 BC), and other sophists of the fifth century, for the fact that they took fees for their teaching. Plato, who was rich enough not to bother, mocked them for lowering the status of the sage.

The new civilization of the Iron Age helped crystallize a new social type[1] in the sophists. The very fact that knowledge of these early thinkers has survived, that Raphael created a masterpiece in their honor,[2] and that legends about their lives has lingered, shows just how important they must have been in the classical world. Irrespective of whether they supported a democratic or aristocratic patron, these thinkers, save the likes of Protagoras, were almost all affluent gentlemen.

Consider this too: the irresistible rise of such philosophers was a global phenomenon. Over much of the developed world, the impact of the Iron Age was strongly felt. In ancient China, thinkers such as Confucius and Lao-tze acted as political or scientific advisers. In early India, there lived at the same time the rishis and buddhas, Siddhārtha Gautama, the Buddha, being the most prominent. And in olden Palestine, the prophets and the subsequent writers of the Wisdom literature, such as Ecclesiastes and the Book of Job, were alive. Indeed, it is thought that Jeremiah may have met the Greek philosopher Thales at Naucratis in Egypt.[3] Many of these thinkers and workers advised assorted princes and attempted in vain to reform their respective governments. But there was one thing they all shared in common: an interest in formulating a worldview of man and nature.

In general, the success of this new social type, the philosopher, is that it filled a vacuum, the gap in ideas left by the shift in the economy from Bronze to Iron Age culture. They provided the knowledge for a new kind of economy. And in that economy, there were new kinds of rulers: princes, tyrants, and merchants. But unlike the great engineers of past regimes who helped build pyramids, temples, and waterways, the philosophers had little to do with the day-to-day running of the economy. They were detached from this material side of government and, as a result, their worldviews were idealist and unsuited to progress in practical science.

But once more, the Greek situation was different; at least in the beginning. Their early sophists failed to fit into the neat picture painted above. At the time of the Ionian philosophers, and the Pythagorean Brotherhood, the slave state and the rule of the rich had not yet fully taken hold.[4] Consequently, the early Greek sophists contrasted starkly with their eastern counterparts. They were at once rational, atheistical, and materialistic, focussed less on ethics, and more on nature.

The Universe, Its Elements, and the Origin of Mankind

In much of the story so far, there has been a necessary focus on developments in astronomy. The reason for this is a good one. The early Greek worldview had inherited a pecking order of priorities. For history shows a distinct progression of the order in which the disciplines of experience were brought within the scope of science. Roughly, it ran mathematics, astronomy, physics, chemistry, and biology.

The order of this development is not easy to explain. But it seems to have been partly conditioned by the practicalities that were in the interest of ruling or rising classes at various times. The control of the calendar, a priestly occupation, led to an initial interest in astronomy, for example. But it was two thousand years before the demands of the rising manufacturers of the eighteenth century gave rise to modern chemistry.

The more complex science of biology, as we shall see later, seems to have been derived directly from the study of its subject matter. Nonetheless, ancient ideas on nature and evolution helped in providing an early general worldview of philosophy and science. So let us take a look at that developing worldview, and the way in which ideas of the origin of life tied in with the developing paradigms.

Consider the Ionians. They came prior to the Pythagoreans and were the first of the Greek sophists. They held that the world was born out of water, and from this aqueous origin came its elements: earth, air, and all living things. It is an account of biogenesis, similar to the creation myth of Sumer, the earliest known civilization in the world, whose first settlement was in the late sixth millennium BC. The Sumerian myth is materialist; it is a reasonable one for people of a delta country where good soil had to be gained from the marshland.

Unlike the version of the same creation myth that appeared in the book of Genesis, the early Greek version was worldly, leaving out the creator. As Protagoras was soon to famously say, "Man is the measure of all things,"[5] and like Laplace just as famously said of God centuries later to Napoleon, "I have no need of that hypothesis."[6] The materialist and atheistic worldview of the Ionians held a fascination with nature, with no need of the metaphysical. For them, all matter was alive. And we will find that this worldview was held by a progression of later philosophers in the same tradition, the Atomists among them.

Anaximander (610–546 BC), Pythagoras's mentor, further refined the Ionian worldview. His creation story summoned the elements of earth, mist, and fire, from which the world was formed. But he also guessed at the genesis of life itself. Fascinated by fossils, Anaximander reasoned that animals long ago sprang out of the sea. The first creatures came to life locked in a bristly bark. As they developed, the bark would dry out, break up. As the planet's humidity dispersed, dry land surfaced and, presently, man was forced to adapt.

This early account of evolution by Anaximander is astonishingly prescient. Censorinus, the third century Roman scholar, gives more detail:

> Anaximander of Miletus considered that from warmed up water and earth emerged either fish or entirely fishlike animals. Inside these animals, men took form and embryos were held prisoners until puberty; only then, after these animals burst open, could men and women come out, now able to feed themselves.[7]

Figuring that the long infancy of humans would make them defenseless in a hostile world, Anaximander thought we would not have survived for long in the primeval world. So he suggested that humans evolved for some time inside the mouths of big fish, protected from the Earth's hostile climate. Only later did they emerge, and lose their scales.

It is clear why some scholars say that Anaximander is evolution's most ancient champion. In particular, his ideas on an aquatic descent of man was resurrected centuries later as the aquatic ape hypothesis. Though he had no theory of natural selection, Anaximander's pre-Darwinian ideas clearly show where we begin to explain the natural world, without recourse to myth or faith. For some this is called the "Greek miracle." But, as we have seen, there is no need of miracle here, any more than miracles are necessary to explain the genesis of life. For us, it is the very beginning of scientific thought, borne out of the material conditions of the time.

Another Ionian, Heraclitus (535–475 BC), is known as the philosopher of change, taking as his motto *panta rhei*, everything flows. Along with his evolutionary idea that everything is in a state of flux, Heraclitus believed fire to be the principal element, since fire's dynamic nature could lead to profound change, transforming all in its path.

Curiously, Heraclitus became known as the "weeping philosopher" because of his melancholia, and as such makes a cameo appearance in Shakespeare's *The Merchant of Venice* (1598). Around thirty years later, Flemish painter Johannes Moreelse depicted Heraclitus, dressed in dark clothing, wringing his hands over the world. This image is perhaps due to the rumor that Heraclitus had a poor opinion of human affairs.

The source of this rumor, as with much of what little we know of Heraclitus, is the biographer of the Greek philosophers, Diogenes Laërtius, who flourished in the first half of the third century AD. With regard to education, Diogenes says that Heraclitus was "marvellous" from childhood, an implication of prodigy, and that though he was born to an aristocratic family, he abdicated a kingship in favor of his brother. Perhaps Heraclitus felt philosophy to be a loftier cause.

His fixation on fire is quite revealing: "All things are an exchange for fire, and fire for all things, even as wares for gold, and gold for wares."[8] Not only does this idea of fire seem like an ancient foreshadowing of the law of the conservation of energy but it also reveals the way in which the sophists based their rather practical philosophy on the material existence of everyday life; in this case, the interchange of ideas between physics and economy.

Heraclitus was the first to introduce a kind of dialectic: the philosophy of opposites. To him, some things—like flame—tend to move up, while others—like stone—tend to move down. This he called the "upward-downward path." Such behavior carries on simultaneously and results in a "hidden harmony."[9] The two opposites were necessary to each other. Since everything flows, and objects are new from one moment to the next, it follows that one can never touch the same object twice. Each object must ebb and flow into regeneration, a continuous harmony between a building up and a tearing down. Heraclitus names these oppositional processes "strife" and suggests that the apparently stable state is a harmony of it, a clear forerunner of the idea of equilibrium.

NATURE AND EVOLUTION

Empedocles (490–430 BC) was another worker in this school of materialist philosophy. He was greatly influenced by the Pythagoreans and considered close in philosophy to the Atomists. But Empedocles is perhaps best known for being the originator of the cosmogenic theory of the four classical elements: earth, air, fire, and water. The elements became the world of phenomena we see today, full of contrasts and oppositions.

By showing through experiment that invisible air was also a material substance, Empedocles first proposed the order of the ancient elements as earth, water, air, and fire. Each element was above the other, striving if disturbed to return to its place in the rightful order of things. He also believed that opposite properties, such as love and hate, were material tendencies that mechanically mixed and separated in a continuous process.

His ideas bear some things in common with the yin and yang dualism of ancient China. Though they probably arose independently, the Chinese also held ideas of two principles, such as fire and water, male and female, forging to form the other elements. In the Chinese version these were metal, wood, earth, and through these by further combinations, the "ten thousand things" of the material world.

The speculations of Empedocles included the idea of a cyclical universe. They are of interest to our story since they also deal with the origin and development of life. In his cosmogony, Empedocles tried to forge a theory of everything. His worldview explains the separation of elements, the formation of earth and sea, of Sun and Moon, and of atmosphere. But he also attempted to account for the biogenesis of plants and animals, and with the physiology of humans. His account looks a little curious to us now, but makes for a fascinating read, and is clearly another crude but quirky anticipation of what has come to be known as Darwin's theory of natural selection.

According to Empedocles, as the elements combined, there first arose strange results: heads without necks, arms with no shoulders.

Next, as these mere scraps of anatomy met, horned heads on human bodies emerged. So, too, developed bodies of oxen with human heads, and creatures of double sex. But largely, these beasts of natural forces perished as suddenly as they arose.

Far more rarely, odds and ends of anatomy adapted to one another, and more complicated creatures carried on. So was the organic universe made, sprung from spontaneous aggregations. The beings produced in this way suited each other, as if this had been intended. In time, various influences reduced the creatures of double sex to a male and a female. The world was replenished with organic life.

We can see the ballpark Darwinism in this cosmogony of Empedocles, based as it is on the combining of his beloved elements. Indeed, his passion for the elements seems to have followed Empedocles to the grave. Diogenes Laërtius reported the legend that Empedocles died by throwing himself into Mount Etna in Sicily, so that people would believe his body had vanished and he had been reincarnated. Alas, the active volcano spewed back one of his bronze sandals, revealing the deceit.

In a comic dialogue written by the second century satirist Lucian of Samosata, Empedocles' final fate is spoofed. Instead of being cooked in the embers of Etna, he is hoisted to the heavens by an eruption. A little singed by his ordeal, Empedocles thrives on the Moon, surviving by drinking dew. And in Matthew Arnold's 1852 poem, *Empedocles on Etna*, an account of the philosopher's last hours before jumping to his death, Empedocles predicts:

> To the elements it came from
> Everything will return:
> Our bodies to earth,
> Our blood to water,
> Heat to fire,
> Breath to air

Fittingly, in 2006, a massive underwater volcano off the coast of Sicily was named Empedocles in the man's honor.

On Change and Constancy

The Ionians view was of a dynamic world, one of constant mutual transformation of its constituent material elements. Put more simply, the mutable elements were responsible for the world in motion. Theirs was a worldview of change, created at a time of great change.

Ironically, their philosophy of opposites was itself subject to its own contradiction. The philosophy had to meet two incompatible roles. On the one hand, it literally stood for the actual material world and its phenomena. Without resorting to gods, their ideas had to account for the daunting ancient landscape of land and sea, sunshine and storm.[10] Indeed, the idea of the fury of the elements lives with us still.

On the other hand, the elements were required to represent quality. An explanation of distinctive characteristics, such as warm and cold, hard and soft, wet and dry, also had to be reckoned within the philosophy's compass. They had not yet nailed each element to a specific material substance, of course; that happened much later with the chemical elements of the nineteenth century. Having said that, Anaxagoras (500–428 BC) did suggest that the seeds of every element were present in all things, a clear ancestor of the idea of states of matter, such as solids, liquids, and gases.

Today, we know that change is key to the cosmos. It is the essential bedrock of evolution. The lapse of ages changes all things: culture, language, the Earth, the bounds of the sea, and science itself. Wherever we look in the natural world, we see systems in motion, from the electron clouds that swarm the heart of the atom, to the gently wheeling galaxies in an expanding spacetime. In society, too, change is the life force. It is the engine that drives progress, the catalyst that sparks the revolution. The only human institution that rejects progress is the cemetery. But change has its enemies. And so it was with the Greeks.

For in great contrast to this progressive philosophy of a material world in flux, is the philosophy that was to come after the Ionians. The majority of philosophers that followed, those in the idealist tradition of Plato and Aristotle, focussed more on stasis. Though they borrowed many ideas from their forerunners, the sophists of later times concentrated far more on the *static natural order* of the elements. Theirs was a fixed and immutable piece of the structure of the universe.

In the hands of Aristotle, this idea of a static order became sanctified, used to constrain any attempts at progressive change, especially social change. In Plato's *Republic*, the aristocracy rule through the "noble lie." Essentially, the pretense was that God had created four classes of men, equated to the elements: the guardians and the philosophers, made of gold; the soldiers, made of silver; and the common people, made of baser metals. Indeed, we still talk of someone "proving their mettle." But the conclusion was clear: the perfect picture of the social world was one where the lower classes were inferior to the upper classes.

This equation, with the natural world on one side, and the social world on the other, merely served to hamper an understanding of either, or both. It twisted a sound, materialist theory into a rigid and abstract one. And it held back for centuries the development of astronomy and science, by burdening philosophy with fanciful notions of a sanctified universal order.

ATOMS AND THE VOID

But it was not all bad news; enter the Atomists.

The great feat of the Ionians had been to create a worldview, one that explained how the cosmos had come to pass, how it worked. Roughly speaking. Though it showed great promise, in that it was a material worldview without recourse to gods and design, its great weakness was in its purely descriptive quality. It needed number. And this injection of number and quantity was exactly what the Pythagoreans provided.

Mathematics was the key. But instead of being tethered to the Pythagorean notion of a universe of ideal numbers, the Atomists imagined a cosmos comprised of tiny countless uncuttable (*a-tomos*) particles. They used the number of the brotherhood, but threw out its mysticism. The *atom* was born. The atoms moved through empty space, their movement describing all observable change. Of course, the theory was not perfect; what theory *is*? They believed atoms to be immutable, unalterable, but we can perhaps forgive them this. They lived at a time before gadgets, before the contraptions of discovery. Nonetheless, the Atomists understood that this particulate quality explained nature's ability to form such rich variety in all things.

The Atomists' next brilliant advance was the *void*: the empty space through which the atoms moved, without artifice. The introduction of the notion of nothingness into science was a bold move. Previous thinkers had mused on the universe with what they regarded as a modicum of common sense. To them, the universe was a plenum, a space entirely replete with stuff. Famously now, we know the idea of a vacuum was abhorred by most prominent philosophers. In conflict with the Atomists, it was Aristotle who was to decree *horror vacui*; nature abhors a vacuum.

Aristotle's reasons were fairly sound. In a complete vacuum, he figured, infinite speed would be achievable. Motion would meet no resistance. To Aristotle the prospect of infinite speed was deplorable, so he decided that a vacuum was equally impossible. But later many of the major triumphs of Renaissance science, such as Galileo's dynamics, were achieved by overthrowing the ideas of Aristotle.

The Atomist worldview also had a radical political side. They stood for a world that took care of itself, with no need of divine dabbling. A world devoid of the preordained. For even though early Atomism was rather deterministic, later exponents allowed for variety and free will in man. We shall soon see that Plato and Aristotle had enough clout to stop the Atomist creed from winning general support. But the creed lived on, and throughout classical times it was a lasting heresy.

One of the foremost thinkers of Atomism was Democritus (460–370 BC). If Heraclitus was the "weeping philosopher," then Democritus was the "laughing philosopher." As early as the first century AD Sotion, the Greek doxographer, or biographer of past philosophers, was already combining the two in a quirky coupling of laughing and weeping philosophers: "Among the wise, instead of anger, Heraclitus was overtaken by tears, Democritus by laughter."[11] And much later, the Italian Renaissance architect Donato Bramante was to paint the fresco "Democritus and Heraclitus," the very painting, housed in the Casa Panigarola in Milan,[12] that introduced this chapter.

Most of the many anecdotes about Democritus speak of his distance, his modesty, and his simplicity. They suggest he lived exclusively for his studies. One tale even has him deliberately blinding himself in order to be less disturbed in his pursuits, though this is likely to refer to the fact that he lost his sight in his dotage. Democritus was highly esteemed by his fellow citizens. Diogenes Laërtius says the reason for such popularity was that Democritus had foretold many things that turned out to be true, perhaps a reference to his vast knowledge of natural phenomena. And of course, Democritus was always eager to see the comical side of life.

Democritus was born in the Ionian colony of Thrace. His father was said to have been so wealthy that he received Xerxes, the King of Persia, on his march through Abdera. To satisfy his thirst for knowledge, Democritus spent his inheritance on his many travels, wandering far and wide through Egypt, India, and Ethiopia.[13] Indeed, Democritus declared that among his contemporaries none had made greater journeys, seen more lands, and met more scholars than himself.[14]

On return to his native land, Democritus traveled throughout the Greek world in order to better understand its culture. Once settled, he occupied himself with writings on natural philosophy and mentions many Greek philosophers in his works, his wealth enabling him to purchase their writings on his travels. But his greatest influence was Leucippus (first half of the fifth century BC), the founder of the Atomism.

Since Atomism is the idea that absolutely anything that exists might ultimately be made up of a collection of tiny discrete parts that cannot be divided further, then it might be applied to even the groupings in society or space. Under his mentor, Leucippus, Democritus seems to have been the first philosopher to realize that the celestial body the Greeks called the Milky Way was actually made up of the light of discrete but distant stars. Aristotle later disagreed, naturally. The Atomists also proposed that life was to be found elsewhere in the universe, a cosmos containing many worlds:

> In some worlds there is no Sun and Moon while in others they are larger than in our world and in others more numerous. In some parts there are more worlds, in others fewer; in some parts they are arising, in others failing. There are some worlds devoid of living creatures or plants or any moisture.[15]

LIFE IN SPACE

The Atomist idea of abundant life in space had begun with the Ionians, particularly Anaximander. This idea of cosmic pluralism, the plurality of worlds, is the belief in numerous other worlds that harbor living extraterrestrials. A universe replete with life. For some, these worlds came and went, appearing and disappearing, some being born as others perished. This movement was eternal, "for without movement, there can be no generation, no destruction."[16]

It is vital to remember that by "other worlds" these ancient thinkers did not envisage today's extrasolar bodies, planets in orbit around distant stars. For the Greeks, remote stars were of little concern. To some they were merely small lights in a vaulted sky that lay beyond the "sphere" of Saturn.

To the Atomists, the host of life-bearing worlds they envisaged was beyond reach. Essentially, they were other "realms," much like the multiverse idea of today, that speculative set of possible universes, invisible and inaccessible from our own Universe. So these supposed other worlds, or other realms, of the Greeks might exist contemporaneously with ours, or form a linear succession in time.

Pluralism met its most fertile form with Epicurus (341–270 BC). Though both his parents were Athenians, Epicurus was born on Samos, the home island of Pythagoras. But he returned to Athens where he founded The Garden, a school named for the garden he owned, and which served as a meeting place with his fellow philosophers. The Epicureans emphasized friendship as a vital ingredient for happiness. Their school appears to have been a moderately ascetic community, quite cosmopolitan by Athenian standards, including women and slaves, and which practiced vegetarianism.[17]

Atomism was more than just a cosmology, of course. But this very early notion of other worlds was very much drawn from the logic of their physics. So it was with the Epicureans. Their idea of an infinite number of worlds was derived from an infinite number of atoms. The "world" of the Epicureans was "kosmos," a system of order, not of chaos. For them, the observable universe was one kosmos. They conjured up the stunning idea of an infinite horde of such worlds, beyond the senses, but not beyond reason. Their Atomist reasoning was acute: there must be an infinite number of atoms, and that an infinite number of atoms could not have been exhausted by our finite world. So, as our world was born by chance collision of atoms in motion, then other worlds must be forged in the same way.

The Atomist worldview would in time spread like wildfire throughout Europe. Its major messenger was the Roman poet Lucretius (99–55 BC). His De Rerum Natura (On the Nature of Things) was an early exercise in science communication, a popularization of the doctrine of the Epicureans. According to Lucretius, other worlds must exist, "since there is illimitable space in every direction, and since seeds innumerable in number and unfathomable in sum are flying about in many ways driven in everlasting movement."

It is an incredible ancient vision of an infinite space, populated by countless stars and planets. Lucretius added that nothing of these worlds was unique, not even the Earth itself, for "when abundant matter is ready, when space is to hand, and no thing and no cause hinders, things must assuredly be done and completed." One thing is very striking about this Atomist pluralism. There is no need of a god. Things are as they are solely due to the mechanism and materialism of their cosmology. It is very unlikely indeed that their concept of infinite worlds would have arisen without the physical principles of the Atomist system and approach.

Interestingly, the atheism of the Epicureans is legend. The word "Apiqoros" appears in ancient Judaism, referring either to those who lack religion, or are seemingly atheistic. The origin of the word is contentious, but is likely to refer to Epicurean beliefs, although it came to imply any philosophy lacking a god. On occasion it is even used to describe heretic ideas or the heretics themselves. The word is still used in modern Hebrew.

ATOMISM: THE SHAPE OF THINGS TO COME

Greek Atomism was not physics. It was not a scientific theory, as such. Its conclusions could not be proven in practice. But without doubt it is the ancestor of all modern atomic theories.[18] Atomism went into a long exile after Aristotle. But one of the early champions of a resurrected Renaissance Atomism was Galileo. Indeed, Galileo's defense of Atomism may have been a major factor in his persecution by the Church.

And around the time Galileo began his explorations with the spyglass, a group of early atomists in England, rather enigmatically known as the Northumberland Circle, were among the scientific vanguard. The Circle held nearly half the confirmed Copernicans prior to 1610, the year of the telescope. But it was French philosopher Pierre Gassendi who was the first of the true modern atomists, drawing his ideas straight from Democritus and Epicurus. In turn, Isaac Newton was a keen atomist, and the ardent inspiration of Newton's theories drove workers such as John Dalton to found the atomic theory of chemistry. The atoms of modern chemistry have not proved as uncuttable as their name suggests, but the more profound principles of nuclear physics lie in the same atomic tradition.[19]

THE DARKNESS RISING

FOOTFALLS OF THE ANCIENTS

In the opulent Apostolic Palace of the Vatican sit the Stanze di Raffaello, the Raphael Rooms. Adorning one of the lavishly decorated walls of this public part of the papal apartments is Raphael's masterpiece, that perfect embodiment of the classical spirit of the High Renaissance, *The School of Athens*.[1]

Painted one hundred years before Galileo's revolutionary discoveries with the spyglass, Raphael's fresco depicts many of the great Greek philosophers from the ancient world, two millennia before. Raphael owes to Michelangelo the painting's expressive energy, its physical power, and the dramatic grouping of his figures.

Raphael was part of the Renaissance cultural movement that began in Tuscany in the fourteenth century. This revival of knowledge, which affected the way in which people viewed and related to the world, was based on such classical sources. *The School of Athens* is a complex painting, reflective of the most learned humanism of the day. But the geometric precision and the spatial grandeur of Raphael's masterpiece leave us in no doubt as to the esteem with which the classical Greek world was held. Raphael portrays a host of Greek philosophers, each in a characteristic pose of activity. The identity of many of the figures is a matter of controversy, but of one thing we can be certain: the painting's two focal figures, standing central amid lofty dome, barrel vault, and the colossal statuary of their fellow philosophers, are Plato and Aristotle.

Now, Raphael hauls us into history. Our journey is to the beginning of the European philosophical tradition. We will discover why philosophy became "a series of footnotes to Plato"[2] and science "a series of footnotes to Aristotle."[3] Our story will show how in due course, and through the hands of the medieval Church, their ideas were used to hold back modern science, including astronomy and biology, for two thousand years.

THE ANCIENT CITY OF ATHENS

Slowly, but surely, the world picture of the Pythagoreans began to corrode.

Its corruption came in the fifth and fourth centuries BC, the central period of Greek thought. And it coincided with two vitally linked developments. One was the dramatic rise of the city-state of Athens, and the Athenian empire. The other development was that in philosophy: the shift from the material to the ideal plane.

The irresistible rise of the city emerged after Greek victory in the Persian Wars in 479 BC. Athens was transformed into the cultural and economic epicenter of the Greek world. With courage and tenacity, they had fought off Persia's imperial aggression. Monetary and military savvy had proved to be priceless. Themistocles, the Athenian soldier and statesman, had persuaded his fellow citizens to plow profits back into the push against the Persians. The coinage used to construct the city's commanding armada came from the Laurion silver mines, one of the chief sources of income for the Athenian state and notorious for the treatment of the slaves who mined it.

The poor powered the navy. This ensured victory for the city, and the support of the common people in its government. For the next hundred years, despite losing the war with Sparta, Athens became the intellectual center of the Greek world. Its ascendancy was such that the literati flocked to the city. Artists and sculptors, philosophers and historians were drawn as scholars to the flame.

Now the change began to tell. The tradition of Ionian science, particularly Pythagorean astronomy, was gifted new impetus. And a period of great significance to the history of science was launched. The imaginative worldview of the Ionians melded with the precision of the new mathematics. Indeed, Anaxagoras, the last of the Ionians, was one of those who settled in Athens. As we briefly related in the last chapter, it was Anaxagoras, friend to Pericles, who was exiled for rationalism in 432 BC.

The main challenges for science were set. Over the succeeding centuries, the natural science of astronomy was used as a means for providing a test of truth, a litmus of wisdom. The task: a philosophy to best explain the motions of the Sun, Moon, and planets. And it is to that task we now turn. Our account will first consider a ponder of philosophers in the Pythagorean tradition: Philolaus (480–385 BC), Heraclides (387–312 BC), and Aristarchus (310–230 BC). Like the philosophers in the previous chapter, these thinkers believed that matter is primary and that number was key to comprehending the cosmos.

Lastly, come the philosophers of reaction. Here we shall mainly consider Plato and Aristotle, and the tradition of the idealist doctrine, which believes that ideas make up either all or a major part of the world of matter. Of course, it is understood that entire schools of philosophy, such as the Stoics, also made important contributions. But for our tale, they are peripheral, as the door swings back from present to past, and we first consider the legacy of the Pythagoreans.

LIGHT BEFORE THE DARKNESS: THE CENTRAL FIRE

Earth became airborne.[4] This first revolutionary step was that of Philolaus, the earliest known thinker to assign motion to our planet.

Philolaus was a Pythagorean. A contemporary of Socrates and Democritus, and junior to Empedocles, he argued that numbers governed the universe. But most significantly for our story, Philolaus is the originator of the theory that the Earth was not the center of the universe. Indeed, his notions of the cosmos are drastically different from what went before. At once, he did away with the idea of fixed direction in space. And he gave science one of the first nongeocentric models of the universe, through his creation of an object known as the Central Fire.

His was a cozy cosmos, one in which you could put your feet up and warm them by the fire. No longer did the Earth stand central, massive, and immobile. Philolaus broke away from the geocentric tradition. At the hub of his "world" was "the hearth of the universe,"[5] The Central Fire. And around this core revolved nine bodies: the Earth, Moon, and Sun, the five planets, and the sphere of the fixed stars. Lastly, and sitting innermost in this snug system, was the *antichton*, or counter-Earth (see Figure 3.1, Schema 1).

The antichton was an invisible planet. Standing between the Earth and the Central Fire the purpose of the antichton was to protect the antipodes from being scorched by the Central Fire.[6] The ancient idea that the western edge of the world was bathed in timeless twilight was now enlightened.[7] It was cast by the shadow of the counter-Earth. Aristotle begged to differ. He scornfully suggested that Philolaus had merely invented the antichton to create a cosmos of ten moving bodies, the magical number of the Pythagorean mystics.

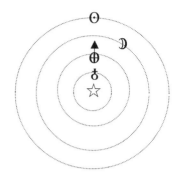

⊕ Earth   ☽ Moon   ☉ Sun   ☆ Central Fire   ⚹ Counter Earth

Figure 3.1   Schema 1: Central Fire cosmology, showing the Earth and Sun at noon. The Earth orbits the Central Fire in twenty-four hours, always turned away from the Central Fire and Counter-Earth. Day lasts as long as the Earth is facing the Sun.

The Central Fire could never be seen. For the civilized and inhabited central region of the globe, the Greek world and its neighbors, was always facing away from the fire (see Figures 3.1 and 3.2, Schemas 1 and 2). Much like the far, dark side of the Moon was always facing away from the Earth. And beyond this cozy cosmos of the Central Fire was its main source of light, the "outer fire."

The "outer fire" bound the cosmos on all sides. A wall of fiery ether, it was an eternal spring of luminous energy. The Sun served merely as its portal or lens, through which the outer light was drawn and dispersed. It is a fantastic notion. But perhaps no more incredible than what we think we know and believe today. Namely, the notion of a ball of burning gas, careering across an endless sky.

Philolaus set free the Earth. Now, the Pythagoreans realized that the daily revolution of the entire sky was merely an illusion, one caused by the Earth's own motion about the Central Fire. And yet, Philolaus did not take the next obvious step. According to some sources,[8] he did not have the Earth rotate on its own axis. Admittedly, his cosmology is innovative enough for ancient times. But having already taken the revolutionary step of releasing the Earth, allowing it to be free and mobile in space, he did not see it spin.

Rather, Philolaus had the Earth revolve, in one day, about his point in space, the Central Fire. By suggesting one complete revolution in twenty-four hours, he enabled an optical illusion. The observer on Earth, now a traveling platform, would see the rich canopy of the sky, turning in the opposite direction. Whether or not the cosmology

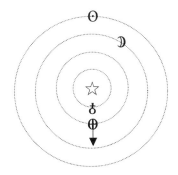

⊕ Earth ☽ Moon ☉ Sun ☆ Central Fire ♁ Counter Earth

Figure 3.2 Schema 2: Central Fire cosmology, showing the Earth and Sun at midnight. The Earth's motion accounts for day and night, which occurs here as the Earth turns away from the Sun, moving around the Central Fire.

of Philolaus included a rotating Earth, this idea of a revolving Earth would see our planet travel and complete a daily orbit of almost six hundred million miles; a fair lick!

The Moon was the sole object in the heavens considered similar to the Earth. Since it was bathed in sunlight for fifteen days at a time, the Moon was thought to be inhabited by life fifteen times as strong as terrestrial creatures. Some Pythagoreans believed the light and dark shadows of the Moon were evidence of the reflections of our oceans. Others thought eclipses were caused by the counter-Earth, whilst some believed the counter-Earth responsible for the soft ashen light of the new Moon. Still others supposed several counter-Earths.

The cosmology of Philolaus sparked a healthy debate in the nervous system of the Pythagoreans. Despite its quirkiness, the elegance and simplicity of his system meant that the idea of a Central Fire quickly became a fixture in the cosmic thoughts of fellow philosophers. As we shall soon see, the ideas of Philolaus were to dramatically influence Aristarchus of Samos. And in time, the Earth's revolution around a Central Fire would become known to astronomers such as Nicolaus Copernicus and Galileo.

A GAME WITH A GLOBE OF GLASS

The Earth was no longer the center of the universe. But the detail of the idea was in doubt.

As salty Greek sailors steered their seafaring vessels, they roamed the ancient Mediterranean, the theater of the Greek maritime empire. Colonies were to be found as far as Iberia in the west to Syria in the east. Towns and cities under Greek control spread from Apollonia on the northern coast of Africa, to Adria, atop the sea that bears its name. So many travels, by so many seafarers, but no sightings of a Central Fire, no news of a counter-Earth. And yet both should have been visible from the "other side" of the Greek world.

At the top of the first chapter we spoke of science as a best current interpretation of the natural world, a cumulative venture. We described its worldview as one that changed with experience. So it was with Pythagorean cosmology. At once flexible and accommodating, the notion of the Central Fire as a spring of luminous warmth was now adapted: the Central Fire moved from outer space, to the core of the Earth itself. The antichton, or counter-Earth, was merely identified as the Moon.

Astronomy as a test of mettle, a test of a philosophy's ability to explain the motions of the heavens. In terms of adaptability, the Pythagorean worldview certainly passed. The next refinement of the model was also to prove crucial.

The Earth was made to spin. No doubt.

Enter Heraclides, the next notable thinker in the Pythagorean tradition. Though, of course, he was not the only worker proposing such ideas at the time, Heraclides is credited by the doxographer Diogenes Laërtius as an important innovator of the cosmic model. Like fellow Pythagoreans Hicetas and Ecphantus, Heraclides understood that the apparent daily motion of the sky was created by the daily rotation of the Earth on its axis.

The innovation also exposed what was to become the central problem for Greek astronomy and cosmology: the annual motions of the seven wandering planets, the Moon, Mercury, Venus, the Sun, Mars, Jupiter, and Saturn.

The galaxy of rotating stars proved no problem. They appeared to never alter their fixed place relative to one another, or to the observer on Earth. As a solid sphere, they rotated en masse, or at least appear to do so, due to the Earth's rotation on its axis.

But in contrast to the seemingly eternal dependability of the sphere of fixed stars were the tramp stars, the planets. Given the relative satisfaction, if not succor, to be gleaned from the regularity of the constellations, how unsettling the wandering stars must have been, especially to philosophers trying to test their wits against the skies. Thank heavens for the Zodiac. At least the tramp stars stuck to this same belt along the firmament.

It is not easy, building a true mental model of the sky. Allow yourself to be transported back to an alternative ancient Greek world, one with a few quirky differences to the days of Heraclides. Your mission, should you chose to accept it, is to prove, with the information available to the Greek naked eye, that our "world" or realm, or solar system if you like, is indeed heliocentric (Sun-centered) rather than geocentric (Earth-centered).

In luxurious comfort, you sit at the center of a transparent Earth, a gargantuan glass globe, clear as crystal. (The comfort is essential, since you need to make painstaking observations). Way above your head is a collection of airborne gods of the Greek world, each flying along a concentric orbit, and each representing one of the wandering planets.

You sit and watch the divine traffic. Each god appears as a luminous point, and each moves at a different speed along a narrow lane, the zodiacal belt. Now the great glass globe begins to move. It rotates around you, as you remain at rest. The direction of the spin of the globe and the flow of the traffic is one and the same. The sphere and the gods rotate in the same direction, but the holy traffic remains confined to the Zodiac.

Let's take a closer look at these orbiting deities. Occupying that part of the Zodiac nearest you, the observer, flies the winged messenger Hermes and the goddess of love Aphrodite, now better known for their Roman names of Mercury and Venus, the "inferior" planets. Slightly above them, and crowned with a shining halo, is the bright white light of Helios, Greek god of the Sun. Higher still fly three gods at different heights: Ares, the god of savage warfare, bloodlust. and slaughter; Zeus, king of the Greek gods; and Cronus, the harvest divinity. In ascending order, these three represent the "superior" planets, Mars, Jupiter, and Saturn.

Saturn is to be found way up in the atmosphere, if great glass globes can be considered to have atmospheres. Above Saturn, there is only the fixed starry sphere, the outer limit of this ancient realm. And that just leaves the Moon goddess, Selene. She is so close to the observer, that her reflection can be seen in the glass surface of the gargantuan globe, for she flies just the other side of the concave wall of the great sphere. Such is the antique geocentric model of the cosmos, complete with soaring gods.

The major problem is this: the system doesn't work. Of course, today the reason is obvious to us. The gods, sorry *planets*, were set out mistakenly. It is the Sun that should be at the center. The Earth should replace the Sun in Figure 3.3, Schema 3, taking the Moon with it. Essentially, the Earth takes the Sun's place between the "inferior" and "superior" planets and we are presented with Figure 3.4, Schema 4.

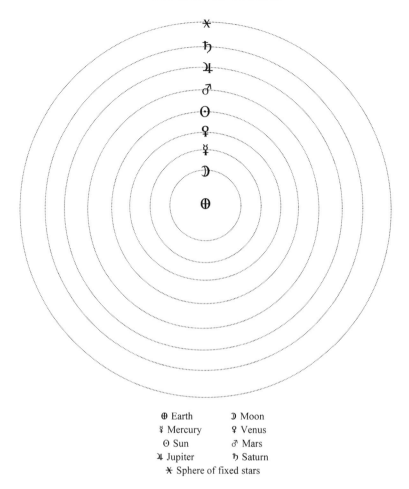

Figure 3.3 Schema 3: The classical geocentric system (with soaring Greek gods!). The planets orbit counter-clockwise, as viewed from above.

So you can see the difficulty. The Greeks were trying to build a picture of the cosmos, from the wrong perspective. And the Achilles' heel of the geocentric model is the baffling behavior of the planets.

Such was the situation facing Heraclides. A philosopher worthy of the name was expected to prove his mettle by clarifying this chaos in the heavens. Looking out once more from the very center of the great glass Earth, the gods of the Sun and Moon are reasonably well behaved. They move smoothly, in a reasonably regular fashion, along the zone of the Zodiac.

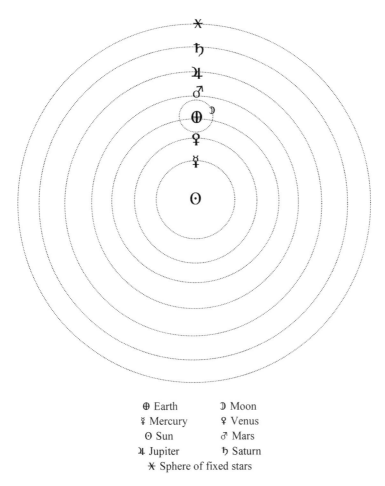

⊕ Earth   ☽ Moon
☿ Mercury   ♀ Venus
☉ Sun   ♂ Mars
♃ Jupiter   ♄ Saturn
✶ Sphere of fixed stars

Figure 3.4 Schema 4: The classical heliocentric system. The planets orbit counter-clockwise, as viewed from above.

Would that the same were true of the other five planetary gods. Alas, they weave their wicked way, ambling along the sky, west to east, on the same bearing as the rest of the heavenly traffic. Then, at intervals, they slow down and stop, and begin winging their way in the opposite direction. As if that is not bad enough, they then seem to have another change of heart, turn around, and resume soaring in the original direction.

And take fair Venus. On some capricious whim, her light seems to periodically wax and wane. This evident flux in luminosity and size seems to suggest that our distance from Venus changes. Looking up through the glass globe, she appears to weave her merry way through the ether. Her behavior is hardly consistent with motion in a circle, with you as the center.

What's more, Mercury and Venus seem to have some kind of unhealthy obsession with the Sun god Helios, forever sticking close, like paparazzi to a princess. True, they may often appear to race ahead, or hang behind, but their general mode is one of pestering. Little wonder that when Venus rose ahead of the Sun, she was known as Phosphoros, the "morning star," and when she set behind the Sun, was known as Hesperos, the "evening star." According to legend, it was Pythagoras who unraveled the mystery; he was the first to realize they were the same planet.

And so to the cosmology of Heraclides. In a bold attempt to throw pure geocentrism into the dustbin of history, Heraclides placed Mercury and Venus in orbit about the Sun. It certainly seemed to explain their erratic movement, relative to the Earth. So he made the two inferior planets satellites of the Sun, and the remainder of the system, Sun included, satellites of the Earth.

Consider Figure 3.5, Schema 5. This compromise cosmology sits somewhere between an Earth-centered and Sun-centered system. The wax and wane of Venusian light is now explained. If she orbits the Sun in the way suggested, then Venus will sometimes approach and sometimes retreat from the Earth. And the strength of her light will change accordingly. The Schema also shows why, as Phosphoros and Hesperos, Venus is to be seen at times ahead of, at others behind, the Sun. And the last piece of the jigsaw slots in, too. The sporadic shifts of both Venus and Mercury are easily described. While the rest of the heavenly traffic is moving anticlockwise, every so often the two inferior planets appear to fly in the opposite direction.

In hindsight, it's so easy to scoff. Admittedly, it's not an accurate model of our immediate universe, but it's quite ingenious. Indeed, this cosmic cunning is evocative of the life of its creator. The son of a wealthy nobleman, Heraclides was packed off to study at the Platonic Academy under the founder, Plato, himself. Though he studied with Aristotle, when Plato briefly left for Sicily in 360 BC, he left his pupils in the charge of Heraclides. Praise indeed, especially when you consider Heraclides was allegedly one of the few philosophers who dared disagree with Plato.

Curiously, the system of Heraclides has a quirky history all of its own. This notion of an Earth-centered cosmos with a heliocentric twist did very well for itself, under the misnomer of the "Egyptian System."[9]

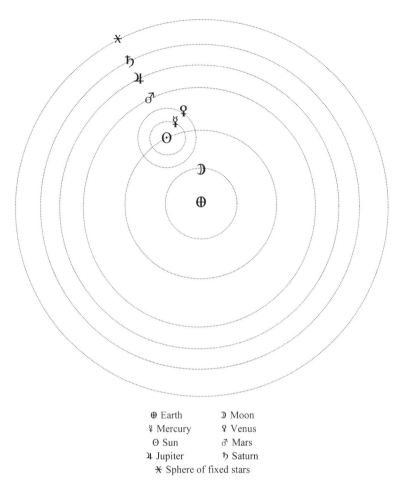

⊕ Earth	☽ Moon
☿ Mercury	♀ Venus
☉ Sun	♂ Mars
♃ Jupiter	♄ Saturn
✳ Sphere of fixed stars	

Figure 3.5   Schema 5: The system of Heraclides. The planets orbit counter-clockwise, as viewed from above.

And it resurfaced many centuries later in a slightly different form in the hands of Danish nobleman and astronomer, Tycho Brahe.

Italian astronomer Giovanni Schiaparelli, he of the Mars debacle many years later, was of the opinion that Heraclides went one step further.[10] Not content with a mere heliocentric twist, it was suggested that Heraclides now had the three "superior" planets, Mars, Jupiter and Saturn, also orbit the Sun. Other scholars went further. They contend that he next set all the planets in solar train, establishing the very first heliocentric system. If he had indeed so done, no trace remains.

SET THE CONTROLS FOR THE HEART OF THE SUN

The final logical step was taken by the last in the line of Pythagorean astronomers, Aristarchus. Hailing from Pythagoras's home island of Samos, this "Greek Copernicus" dared to put the Sun and not the Earth at the center of the universe.

Alas, the details are lost. The original text in which Aristarchus set out his system of the world is no longer extant. Nevertheless, his references are impressive. Copernicus himself was to resurrect eighteen centuries later this pinnacle of Pythagorean astronomy. And we have the testimonies of Plutarch and Archimedes that Aristarchus advanced the alternative model of a heliocentric system.

Archimedes was one of the leading thinkers of antiquity. Mathematician, engineer and astronomer, he was a younger contemporary of Aristarchus. One of Archimedes' most remarkable works is a short treatise called *The Sand Reckoner*, in which he tries to determine an upper limit for the number of grains of sand that the universe can hold. A rather quirky idea for a book. It is about only eight pages long in translation and is dedicated to King Gelon II of Syracuse. Crucially, to meet the challenge he set himself, Archimedes used the system of Aristarchus:

> You King Gelon are aware the "universe" is the name given by most astronomers to the sphere the centre of which is the centre of the Earth, while its radius is equal to the straight line between the centre of the Sun and the centre of the Earth. This is the common account as you have heard from astronomers.
>
> But Aristarchus has brought out a book consisting of certain hypotheses, wherein it appears, as a consequence of the assumptions made, that the universe is many times greater than the "universe" just mentioned. His hypotheses are that the fixed stars and the Sun remain unmoved, that the Earth revolves about the Sun on the circumference of a circle, the Sun lying in the middle of the orbit, and that the sphere of fixed stars, situated about the same centre as the Sun, is so great that the circle in which he supposes the Earth to revolve bears such a proportion to the distance of the fixed stars as the centre of the sphere bears to its surface.[11]

The other prominent reference to the system of Aristarchus is that of the Greek historian and essayist Plutarch. In his *On the Face in the Moon Disc*, Plutarch presents a long and curious treatise on topics such as the cause of the Moon's light, its peculiar color, and the possibility of its being inhabited. The dialogue refers to Aristarchus of Samos who believed, "that the heaven is at rest, but that the Earth revolves in an oblique orbit, while it also rotates about its own axis."[12]

Aristarchus was a materialist. His sole surviving work is a short treatise, *On the Sizes and Distances of the Sun and Moon*. And it is stunning. We spoke in Chapter 1 of science as more than a mere matter of thought. Rather, it is the process of thought carried into action. So it was with Aristarchus. This Pythagorean was not content with just contemplating the cosmos. He wanted to get a real measure of it. His treatise presents a way of calculating not just the sizes of the Sun and Moon but also their distances from the Earth, in Earth radii.

Aristarchus began with the half-lit Moon.

He reasoned that when the Moon was exactly semi-lit, quadrature, it forms a right-angled triangle with the Sun and Earth. Indeed, quadrature occurs when any two heavenly bodies appear 90 degrees apart from one another as viewed from a third body. Aristarchus realized that, by measuring one of the other angles in this right-angled triangle, he could work out the ratio of the distances to the Sun and Moon using trigonometry.

Simply brilliant. But Aristarchus didn't stop there. He found that the distance to the Sun was around twenty times greater than the distance to the Moon. And since the apparent sizes of the Sun and the Moon in the sky are the same, the Sun must be twenty times bigger. The method is sound. It is the measurements that proved tricky.

He then turned his attention to eclipses. To find the actual sizes of the Sun and Moon, Aristarchus used two observations. First, that the disk of the Moon just covers the Sun, during a solar eclipse. And second, that during a lunar eclipse, the Earth's shadow appears thrice as large as the Moon, at the Moon's distance. Today, using his eclipse schema, it is possible to show that the Sun is about four hundred times farther from the Earth than the Moon, and its diameter is around one hundred and ten times greater than that of Earth.

His book is a classic of antiquity. The image we get is of a contemporary and rational worker, inventive in ideas and precise in practice. Astronomers used the way Aristarchus derived the distance of the Sun for centuries to come. His elegant method is good. If the results are in error, it is only because he was working two thousand years before the invention of the spyglass and Galileo.

So this brilliant train of cosmic thought that began with Pythagoras, reached its heady height with Aristarchus. A clear eighteen centuries before a Copernican dawn came to the scientific revolution, Aristarchus imagined a Sun-centered universe. And it is no mere guesswork. The cosmos of Aristarchus sits clearly in the Pythagorean tradition. One that saw Anaximander set the Earth adrift, Philolaus warm the skies with his Central Fire, and Heraclides conjure up a cosmos with a heliocentric twist. Markers on the road to a Sun-centered system.

And yet the system of Aristarchus did not prevail. It was not the materialist approach that won out in the bid to crack the puzzle of the planets. Roman engineer and writer Vitruvius said of Aristarchus that he was one of the leading astronomers of his time, a universal genius. We know better. For Aristarchus stands on the shoulders of fellow philosophers in the Pythagorean tradition. And in spite of all this, science now began to fall into disgrace and decay.

THE PHILOSOPHY OF REACTION

Athens fell.

The glory that was the ancient Greek world began to lose its luster. The city-state of Athens had arisen with a planned and real-ized citizen democracy, the first in history. It was a political system that showed huge creative and cultural promise. But it didn't last. It finally fell because it was founded on slavery and the exploitation of lands within the Athenian empire. It became more prone to the blitz of aristocratic forces, made flesh by the shock troops of Sparta, backed by Persian gold.

It was a decisive moment in the classical world. Athenian democracy had shown how the rule of the wealthy could be challenged by a more popular control of social life.[13] But the collapse of democracy spelt ruin for Athens and the city-state. Now, despite material and cultural growth, Athens was doomed. Democracy had provided a way out of the contradictions of the Iron Age city. With democracy dead, the only other option was slavery at home and war abroad. So was the civiliza-tion of the Greeks spread out over a great part of the globe. But at heart, it was fading.

Plato and Aristotle belonged to this Athens, the Athens of decline. They derived their awesome capacity to control classical thought from the revolutionary nature of the first free city. But their philosophy was one of reaction. They poured scorn on democracy, hardly managing to conceal their deep dread of such freedom.

It was a time of great change. On the face of it, it is extremely implausible to deny change, but extreme implausibility has not always worried philosophers. The Eleatic school had been the first to do so, especially Parmenides. Allied to the aristocracy, they had violently attacked experimental science. Observation, they claimed, was merely opinion, fallible to the vagaries of the senses. Truth was not to be found in the blind eye, the echoing ear. Change is impossible, they claimed: the real universe is whole and immutable. So speaks the deep desire for fixity that rears its ugly head, on the losing side, in times of trouble.[14] It is the mantra of the desperate.

Now Plato mouthed that same mantra. His task: to stop the world from changing. Greek science had always had a different nature from the civilizations that went before. It was not only more rational, but also more abstract. A universe that can be revealed by pure logic served Greek science well, initially. It freed men from their superstition. But now it was to prove ruinous. Under the influence of Plato and Aristotle a more abstract and a priori approach managed to convince generations of thinkers into believing they had solved problems before they had barely begun.

Science was separated from technique. Traditionally, as we discussed before, science is based on solving technical challenges or easing administrative tasks. Greek science, however, was based on general principles, somewhat removed from technique. The mathematics and astronomy the Greeks valued so highly were based on deduction and proof. But the immense prestige of these disciplines meant that the scholarly applied principles in general that only actually worked for a limited part of the natural world. And even then only where the grunt work of observation had been done. Witness the work of Aristarchus and his meticulous observations of the Sun and Moon.

Even more so than ancient Egypt and Mesopotamia, the Greek hand worker was considered inferior to the brain worker or contemplative thinker, the cogitator. This social class division was greatly buttressed by the association of hand work with slavery, particularly in later Greek society. Free men did much technique, of course. But they were nonetheless degraded by comparison with slaves, their work described as base or servile.[15]

So, despite the fact that the philosophers derived some of their conclusions as to how nature behaved from the work of craftsmen, they rarely had experience of that work. What is more, they were seldom inclined to improve it, and so were powerless to pry apart its potential treasure of knowledge that was to lead to the scientific revolution in the Renaissance.

PLATO THE IDEALIST

The separation of science from technique met its highest form with Plato. An affluent young aristocrat from Athens, he famously became a student of Socrates—but in the Athens of decline. Plato's political aspirations seemed forever fated by the return of democracy. Perhaps as much influenced by the thinking of Socrates as by what he saw as his teacher's unjust death, Plato resolved to commit his life to philosophy. His goal was to lead men to a better life, by discovering the values and ideology of a perfect State.

His aim led Plato to idealism, that branch of philosophy on the side of the social order more often than not loyal to the aristocracy and established religion. As has already been touched upon in the story so far, a careful study of the history of science reveals a clear characteristic: the struggle of two main conflicting trends. Since the very dawn of science, there has been one tendency toward a formal and idealistic approach and another tendency toward a more practical and materialistic approach. Though it very probably originated with the formation of the first societies, this conflict was a main feature of Greek science.

Plato was a master exponent of idealism.

His philosophy reads like a black amalgam of Aldous Huxley's *Brave New World* and George Orwell's *Nineteen Eighty-Four*. We have already spoken of Plato's "noble lie": the pretense that God had made four kinds of men. Like all idealists, the aim of his philosophy and science was to describe the way of the world, and to show how hopeless it was to try changing its fundamentals. For Plato, it merely remained to dispense with democracy, and rule the republic through its true guardians, the men of gold. This ideal form of government may not be readily apparent to inferior classes. So it was necessary to show them the illusive nature of reality, and the unreality of evil within it.

In works such as *The Republic* and *The Laws*, Plato set down a constitution. It affirmed the right of the best people, the aristocracy, to rule in perpetuity. And it sought a way in which this state of affairs could be made acceptable to the lower orders. Thus the division of the citizens of the Republic into four classes: the guardians, who governed; the philosophers, who ruled; the soldiers, who defended; and the people, who did all the work.[16] Each class was made of a certain metal—gold, silver, brass, and iron—and associated with a particular humour—yellow, white, red, and black. With this unyielding system, Plato hoped to gift Athens a stable government.

A little leeway was allowed. A few gifted members of the lower classes were let into the upper classes. It was a safe way of maintaining their rule.

> If the rulers find a child of their own whose metal is alloyed with iron or brass, they must, without the least pity, assign him the station proper to his nature, and thrust him out among the craftsmen and farmers. If, on the contrary, these classes produce a child with gold or silver in

his composition, they will promote him, according to his value, to be a Guardian.[17]

A slew of other pious deceits was advised. To improve the Greek race, marriage would be abolished and lots drawn. But they would be secretly rigged, according to the principles of Plato's eugenics. Strict censorship would also be enforced. The youth would not be allowed to read Homer. For Homer's works not only inspired disrespect for the gods they also tempted a fear of death. No point in discouraging the lower orders from dying in battle.

The beauty of Plato's prose, it seems, blinded readers to the unsightly nature of his ideas. In the words of Welsh philosopher Bertrand Russell, "That Plato's *Republic* should have been admired, on its political side, by decent people, is perhaps the most astonishing example of literary snobbery in all history."[18]

Platonic Astronomy

The conflicting contrast of the idealist and materialist approaches now becomes stark. The materialist view of the cosmos that infused the astronomy of the Pythagoreans and the Atomists was a practical one. Crucially, it is a philosophy of matter in motion, an account of nature and society from below and not above.[19] A potent example of its power is to be found in Lucretius's Atomist verse *On The Nature of Things*. It broadcasts the boundless energy of a world in flux, and man's power to alter it, by learning its laws. Applied to society, you can see its political threat to the established order.

Later, when we look at developments during the Renaissance, we shall see how the experimental science of Galileo kicked against its main enemy, the idealism of Aristotle backed by the Church. And in the nineteenth century, this rolling contradiction is to be found riddled through the enmity between religion on the one hand, and the science of the "Darwinian revolution" on the other.

In many ways, the continuation of this conflict hints at something deeper than just the science at stake. Rather, it is a reflection of an economic and political struggle. During each age, idealism is used to pretend that social problems are illusive. And at each age, materialism has invoked practical tests of the real world and insisted upon the need for change.

So it was with the Greeks. The idealist science of Plato was in many ways a response to the materialism of the Atomists. Plato's politics saturated his science. In both his philosophy and his physics, change was evil. Ideal, truth, and beauty, all were eternal. And beyond question. Since they were not to be found here on Earth, they remained to be realized in an immutable heaven.

For Plato, the beauty of the stars was just one facet of the visible world. In "truth," they were a diffuse copy of the real world of ideas. And any attempt to understand them, or to calculate their motions, was therefore absurd. In Plato's words, "Let us concentrate on (abstract) problems, said I, in astronomy as in geometry, and dismiss the heavenly bodies, if we intend truly to apprehend astronomy."[20]

It is the greatest of Greek tragedies. Along the materialist road, there was only one stop from Aristarchus to Copernicus, one stop from Archimedes to Galileo, and only one stop from the Atomist creed to Darwinian evolution. And yet it was Plato and Aristotle that reigned supreme, and their robust defense of a system that was already in ruin.

A hundred years of Greek civil war had made exiles of its politicians. Chronic civil unrest had led to political and moral insolvency; economic hardship had made homeless many of its citizens. The classical Greek world was facing a stark crisis. In many aspects, scholars regard it as one of the greatest failures in history.

> Plato and Aristotle each in his different way tries (by suggesting forms of constitution other than those under which the race had fallen into political decadence) to rescue the Greek world, which was so much to him from the political and social disaster to which it is rushing. But the Greek world was past saving.[21]

The crisis that faced the Greek world bled into its physics. Plato invented astrology. The word, rather ironically, means reasoning (*logos*) about the stars, the old "astronomy" had merely meant to order (*nomos*) them. Plato developed the mystical views of the Pythagoreans, especially with regard to the cosmic significance of mathematics. But it was a consciously abstract approach. And it took number away from its practical origin and development, toward a mere contemplative plane.

The ancient popular view of the heavenly bodies was that they were divine beings. This was especially true of the Sun, Moon, and planets. The belief explains why old-fashioned folk were shocked by the impious Ionians, who suggested the planets were merely fiery globes wandering through an empty sky. Heaven forbid.

Plato saved the day. But the cost to science was severe. He united mathematics and theology and declared that the planets were divine.

Their divinity shone through in their unchanging and regular paths, orbits of perfect and circular movement. Thus was the inaudible harmony of the spheres made. Just as a finger gently rubbed along the rim makes a wine glass sing, so the divine planets sang in their circular orbits about the central Earth.

All this despite the fact that evidence already existed to the contrary. Indeed, a Hippocratic contemporary of Plato's opined, in relation to the "mysterious" illness of epilepsy, "It seems to me that the disease is no more 'divine' than any other. It has a natural cause, just as other diseases have. Men think it divine merely because they do not understand it. But if they called everything divine which they do not understand, why, there would be no end of 'divine' things!"[22]

And yet Plato said the stars were divine. Change was cast out of the heavens, as he no doubt would have cast it out of society. The philosopher's highest calling became the consideration of perpetuity, his supreme pursuit the proof of man's immortality. And for science, long term oblivion. The challenge that materialist philosophy had presented to faith now crumbled. The world of matter in motion, proposed by Pythagoreans and Atomists alike, ground to a celestial halt.

In pursuit of heavenly perfection, Plato had concluded that the shape of the world was a perfect sphere, and all motion must be in perfect circles at uniform speed. And the mission he now passed on to mathematicians was to imagine a cosmic model. One in which the evident irregularities of the paths of the planets across the Greek sky was explained away by a system of regular motions in perfect circles.

Platonic astronomy is a strange contradiction. His contribution to the field was trifling and yet his influence was vast. His interest in the subject was minimal, and yet together with Aristotle, he held back science for two thousand years. Any valid physics of the stars was now held in stasis until the days of Copernicus and Galileo. That Plato's authority should have been so enduring is explained by a number of influences. But two factors stand head and shoulders above the others: one is The Academy, the other is Aristotle.

PLATO'S ACADEMY

As a wealthy Athenian aristocrat, Plato's lineage had served him well. He had traveled in the hope of getting his political views established through a prince who was also a philosopher. His final effort in this regard was with the tyrant, Dionysius the younger of Syracuse. But it became clear that the young prince could not stand the mathematical rigor Plato thought necessary. So he finally returned to Athens, being seized and almost sold as a slave on the way home.

Once back at Athens, Plato set up the first institution of higher learning in the Western world. Here, for almost half a century, he taught at the academy, named after the hero Academus. Plato defined the character of the institution, directed the tone of its debates. Above the gate, he had written the legend "Let No One Ignorant of Mathematics Enter Here." The goal was pure knowledge, mainly astronomy, mathematics, and music. But it was abstract learning, plagiarized by the reading of texts; not wisdom derived from a close study of nature, with all its twists, turns, and trickery.

Within these walls, Plato produced an immense body of work. He became the first philosopher of antiquity whose writings survived, not in pieces of parchment, but in sheer bulk. The extent of Plato's work is vast. Its scope covers a great part of the ancient sphere of knowledge; his dialogues alone comprise a volume the length of the Bible.[23]

Here lies the first answer to Plato's enormous influence. The academy is the distant ancestor of all of today's universities, seats of learning, and societies of science. The teaching did not stop with Plato's death. For almost one thousand years the Athenian Academy continued, poring over the works of Plato, seeking knowledge of truth and beauty for its own sake. Until it was closed in 529 AD, the fellowship of the academy passed on Plato's mystical idealism. A static philosophy, preserved in amber.

ARISTOTLE'S UNIVERSE

The universe of the ancients had been a kind of cosmic oyster, clammed up in space and time. The Pythagoreans had forced it open, let a spherical Earth adrift, as the Atomists inflated its outer limits beyond infinity. It was a cosmos of change, a whole "world" in motion. Aristotle wound back the cosmic film, returned an immobile Earth to its center, and popped the genie back in the bottle.

The Aristotelian universe. The closed cosmos resurrected with hard limits in time and space.

For two millennia his cosmology held sway. Aristotle's vision was of a two-tier, geocentric cosmos. The Earth, mutable and corruptible, was placed at the center of a nested system of crystalline celestial spheres, from the sublunary to the sphere of the fixed stars. The sublunary sphere, essentially from the Moon to the Earth, was alone in being subject to the horrors of change, death, and decay.

Beyond the Moon, the supralunary or celestial sphere, all was immutable and perfect. Crucially, the Earth was not just a physical center. It was also the center of motion, and everything in the cosmos moved with respect to this single center. Aristotle declared that if there were more than one world, more than just a single center, elements such as earth and fire would have more than one natural place toward which to move, in his view a rational and natural contradiction. Aristotle concluded that the Earth was unique.

Let's unravel this model of Aristotle's a little more, peel back the layers, and peek inside. After all, this is the system that was to stand the test of time. A synthesis of earlier Earth-centered cosmologies, it would last until the scientific revolution and have an immense influence on medieval religion and culture.

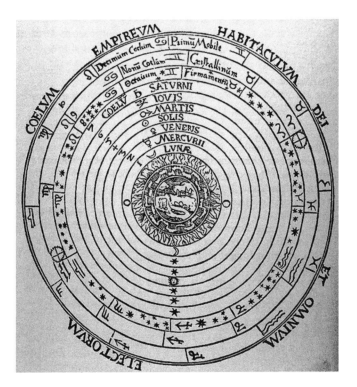

Figure 3.6 Aristotle's geocentric universe, complete with the four elements, celestial spheres and the Primum Mobile, the ninth sphere driven by divinity. Beyond lay heaven, 'the habitation of God and the Elect'.
Source: The illustration is from Peter Apian's *Cosmographia*, Antwerp, 1539.

Aristotle's walled-in universe encases nine concentric spheres, clear as crystal. The outermost sphere is that of God, the Prime Mover. It is God who spins the world from outside. The motion He imparts to the outermost sphere is transferred to each adjoining sphere in turn. Like a child's clockwork toy, the sky is reduced to a mechanical curiosity, God himself keeping the machinery in motion.

In the Pythagorean version, Philolaus had filled the entire cosmos with a source of cosmic energy, the Central Fire. With God's removal to the outer limits, Aristotle now bestows upon the Earth the most lowly and humble place in the whole universe, the central region. It is only this innermost layer, the sublunary sphere, which is privy to dreadful change.

Beyond the sphere of the Moon, all is serene. This supralunary region houses each of the planets in their God-induced motion about the Earth. It is the classic geocentric system of the ancestors, dancing to Plato's divine tune of regular motion in a perfect circle. To the ancient Greeks, the system gifted cosmic comfort to a frightened world. And later to medieval mind, the division of the cosmos into two, one part lowly, the other divine, one part flux, the other eternal, brought the illusion of strength in times of turmoil.

Caught in a crossfire of cosmologies, Aristotle's other gift is that of supreme compromise. His universe is a union of worldviews, a meeting place of materialism and idealism, and a profound consolation to the minds of the meek.

His sublunary sphere is the domain of the materialists. It is the region of Aristotle's universe that is the result of Heraclitus's dynamic forces in constant flux. Here, matter is made up of various fusions of the four classical elements of Empedocles, earth, water, air, and fire. Each has a natural place to be, earth downward, fire upward, air and water horizontally.

The four elements are agents of change. They forever transmute, the sublunary atmosphere replete with their fusions. Indeed, the makeup of the atmosphere is not pure air. Rather, its substance is a catalyst of change, which when set in train ignites to create meteors and comets. In short, the sublunary sphere is buzzing.

The rest of his universe is dormant, divine but essentially dull. This is the domain of the idealists. The influence of Parmenides on Plato and Aristotle is telling here. His teaching that change is an illusion finds form in Aristotle's planetary orbits, his idea of static perfection is writ large in the outer limits of Aristotle's cosmos.

For beyond the Moon there is no change. The four terrestrial elements that provide the key to change in the sublunary sphere are absent here in the supra-lunary domain. True, the heavenly bodies move in orbit about the Earth, but it is motion without change. For circular motion is perpetual, without beginning or end, one continuous movement that returns into itself. Parmenides realized.

The fabric of the cosmos was quintessence, the fifth element. Pure and unchanging, its natural circular motion explained the planetary paths. Its appearance in the shape of crystal concentric orbs is flawlessly realized, for the sphere is the only perfect form. And the further we fly out, from the Moon to Mercury and beyond, the purer the quintessence becomes. Until it meets its highest form in the sphere of Aristotle's God, the Prime Mover.

THE GREAT CHAIN OF BEING

Our account of Aristotle's two-tier universe is a schematic one. The portrayal of one tier where "everything flows," and another tier where "nothing ever changes," is intentional. It signifies a period of turmoil in the Greek world, a time of great debate. The lives of Aristotle, Epicurus the great Atomist, and Zeno the founder of the Stoics all crossed. So, our point is to give a distilled essence of that debate, and the powerful way in which Aristotle's model represented a careful compromise of divergent viewpoints.

As we shall see in the following chapters, the Aristotelian universe was to prove crucial to the Copernican and the Darwinian revolutions. Its potency became stronger still when its two-tier simplicity became the multitiered cosmos of the Latin medieval world. The system that emerged was the *scala naturæ*, the Great Chain of Being. A grand scheme of the universe, its main characteristic was a strict hierarchy, in which every object and creature had its exact assigned place.

Aristotle's stance on change was far less damning than Plato's, and it was more ambivalent. Though he rejected evolution and progress, Aristotle nonetheless regarded all change in nature as purposeful. As an eager biologist, he saw natural movement as a kind of object-oriented process; even inanimate bodies were goal-driven. Just as a salmon swims upstream, so a stone will fall to earth. Just as a spider weaves its web, so the flame of a fire will reach for the sky. Each had their "natural place" in the grand scheme of things.

Plato's abhorrence of change resonated with the Church's notion of an immutable heaven. He had rendered a lowly Earth, set in a walled-in universe of nested spheres, shielded from uncivilized change. Aristotle's elaboration was to have a devastating effect on the course of Western science, the *scala naturæ* gifting the Church an ornate system of control—one does not abandon one's place in the chain; it is not only unthinkable, but generally impossible.

During the long centuries ahead, the philosophy of Plato and the physics of Aristotle were to have a massive stranglehold on development. They developed a system that, though contrasting in many respects, taken together seemed to present a complete solution to the troubles of their time. As the cultural climate evolved in Europe, Plato and Aristotle bestowed bedrock upon the ideological superstructure that was built by the medieval Church.

PART II

THE GATHERING STORM

CHAPTER 4

MEDIEVAL SKY

Dante Alighieri's epic poem the *Divine Comedy* is stunningly realized by Domenico di Michelino's painting, *Dante and The Three Kingdoms.* Completed in 1465, the painting shows Dante, holding his book aloft and surrounded by the three realms: Purgatory behind him, the heavenly City at his left, and to his right a procession of tortured souls descend into the circles of Hell, to which Dante gestures.

Dante's work describes the poet's epic journey through the fourteenth century Christian universe.

His adventure begins on the surface of the spherical Earth. Into the Earth he then descends, through the nine circles of Hell, which mirror the nine spheres of heaven in Michelino's picture. Dante passes through the sixth circle of hell, which contains the flaming tombs of the ungodly Epicureans. At last, Dante reaches his first destination, the most corrupt of all realms, the squalid center of the universe, locus of the Devil and his legions.[1]

The poet then surfaces at the other side of the world, where he finds the Mount of Purgatory, shown in the painting with its base on the Earth and its peak, complete with Adam and Eve in the Garden of Eden, reaching into to the aerial regions above. Passing through purgatory, Dante now soars through the heavens, first passing through the terrestrial spheres of air and fire, then flying through each celestial sphere, speaking with the spirits that inhabit them. Finally, he approaches the Primum Mobile, the first moved thing and the sphere that imparts motion to the rest of the realm. Once there, Dante beholds this last sphere, the Empyrean, God's Throne.

The universe of Dante's *Divine Comedy,* and Michelino's portrait, is Aristotle's, tailored to the medieval Holy Church and to God himself.

It is a universe transformed through Christian symbolism. By his use of allegory, Dante strongly states that the medieval universe could have no other structure than the Aristotelian. In his powerful poem, a masterpiece of world literature, Aristotle's universe of spheres mirrors both Man's hope and his fate.[2]

Both bodily and spiritually, Man sits midway. His crucial position in the hierarchical chain is halfway between the inert clay of the Earth's core and the divine spirit of the Empyrean. The rest of the universe is made of either matter or spirit, but uniquely Man is made of both, body and soul.

Man's place is also transitional. He lives on Earth, in filth and uncertainty, close to Hell. But he is at all times and in all places under the all-seeing eye of God, and with full knowledge of the heavenly spheres above. Man's dual nature and place in the great scheme of things helped impose the dramatic choice that faced all Christians in the medieval world. Whether to follow his base and human nature down to its natural place in Hell, or to engage with his spirit, and follow his soul up through the celestial spheres to God.

This epic poem, *The Divine Comedy*, is "the vastest of all themes, the theme of human sin and salvation, is adjusted to the great plan of the universe."[3] The adjustment had been achieved. The western world lay at the feet of the medieval Holy Church. Any alteration in the great plan of Aristotle's universe was bound to corrupt the drama of Christian life and death.

To shift the Earth was to sever the continuous chain of created being, and to move the Throne of God himself.

THE LEGACY OF THE GREEKS

Our concern in this book is to show that scientific revolutions are not about "Great Men." They are about movements of economy and society, politics and people, played out on a global scale. In particular, the first section of the book focussed on the life and times of the Greek classical world, especially in relation to the natural sciences of astronomy and biology, such as it was. Our purpose now is to look at the way in which the intellectual and cultural brilliance of the Greeks greatly influenced succeeding ages. We shall not be disappointed.

And yet despite this cultural brilliance in theory, surprisingly little was translated into practice, especially when compared to the developments of later ages. The benefit of hindsight is a wonderful thing. Nonetheless, their knowledge and skill may have been realized in the trappings of grandeur of the ancient Greek world: the scale of their cities, the beauty of their temples and pottery, and so on. But the way of life for the vast majority of people in the so-called civilized world was no better than it had been two thousand years before.

The separation of science from technique had resulted in little progress in the material reality of daily life. Sure, there was great sophistication in their mathematics and philosophy. But this should not blind us to the fact that housing, farming, and clothing had seen scant improvement. The science of the Greeks had met with only modest application.

It is little wonder, really. Science was in the hands of an elite, a narrow class of well-off citizens who served their own interests. They despised technique, and the qualitative science they practiced was far too limited to be of much practical use. Their mathematics was elegant, but their mechanics inexact, their astronomy magnificent, but their maps indifferent.

The picture was bleaker still for the other natural sciences. Naïve or mystical explanations, founded on narrow frameworks such as the elements or the humors, meant that a true experimental understanding of nature was elusive. Turgid and discursive journals were the norm. The only extant work of the ancient naturalist Pliny the Elder is a case in point. His *Naturalis Historia* was merely a compilation of observations by cooks, doctors, fishermen, and farmers.

In short, the full potential of Greek science could not be realized. The cradle of the culture that gave it birth was rife with restrictions. The economic and social constraints of a slave-owning plutocracy meant that the day of Greek science was yet to come. How else are we to explain the Greek tragedy? There is only one small step from Aristarchus to Galileo, but a world of difference in economies. One economy powerless to deliver from disaster, the other able to conjure up a spyglass, and a scientific Renaissance.

Through sheer weight of authority, and the potency of its theory, the Greek contribution to future generations of philosophers was fundamental. Though their culture did not have enough power to save itself, it had enough prestige to be revered and later act as a catalyst to spark a revolution.

The most crucial treasure of the classical world was the very notion of natural philosophy.

Indeed, the legend of the ancients went so deep that their reputation was greatly exaggerated, the stories wildly inflated. It was said that the Greek study of nature was so complete that they were able to control it. Alexander the Great, for instance, was said to sail the seas in a submarine, built under the instruction of Aristotle. The great philosopher was also told to have been instrumental in the building of Alexander's chariot, which could take to the air, powered by eagle.

Astronomy was the most enduring element of Greek culture. It provided the blueprint for Dante's depiction of Christian life and death, the architecture of the universe of the medieval Holy Church. And most of all it was a vital aid to navigation, the plotting of the planets essentially charting the voyages of discovery. Astronomy had a cash value.[4]

As for the other sciences, they were suspended in text, lying dormant and dusty until the Arabs and Renaissance scholars rediscovered them. Historians will never know what wisdom was lost, what nuggets of knowledge are irretrievable. But a strong enough signal survived to shine through and inspire the thought and practice of the modern age. The greatest and most lasting legacy of the Greeks was in their science, a force we feel to this day, and one that was rediscovered in the medieval world.

THE AGE OF FAITHS

To trace the tale of two scientific revolutions, we need to know the backstory. To explain the birth of modern science, we need to know how it happened at all. What led to the emergence of the new science in Galileo's Italy of the sixteenth century, and why had it flourished so fantastically in Darwin's England of the nineteenth century?

The most important dynamic on the world stage that led to these developments in science was the economic factor. The period of time we deal with in this chapter is vast, ranging from the end of the classical world in the fifth century to the birth of the scientific revolution, based on a new economic system and a new experimental science.

Indeed, it is a rather long and tortured birth, ten whole centuries that spread from a slave-owning plutocracy through to a feudal world. But increasingly, throughout the later Middle Ages, a premium was placed on the development of labor-saving devices, time-shaving technology. These same technologies were accountable for the change from a feudal economy into an emergent capitalism. Science at first shadowed the rise of nascent capitalism; later, it came to dominate its development.

History's obsession with the Roman Empire has traditionally dictated a rather false view. One that portrays a collapse of civilization, between the third and ninth centuries. In truth, the reality was different. The collapse occurred only in those tardily and allegedly civilized parts of the ancient world, including Britain, France, Rhineland, Iberia, and Italy. In these lands, a system of government disintegrated—a class system, run by wealthy slave owners.

In the meantime, the rest of the world carried on regardless. The great cities of Alexandria and Constantinople survived intact. The territories of China, India, and Persia continued to flourish and develop unabated.

So much for the Dark Ages.

In the Latin West, however, the lands recently deprived of the civilizing influence of slavery, merely moved on to a more widely based feudal order. Indeed, and in great contrast to the slave system that preceded it, the economy of the whole period from the fifth to the seventeenth century may be taken as feudal.

It was an economy based on the country. The economic unit was the manor, villa, or estate, worked by serfs or peasants, rather than chattel slaves. These serfs remained in possession of their tools and land, but were forced to give up part of the produce or labor to their lords in the form of taxes, rent, or feudal service.[5]

Trade was at a low level; the land producing practically everything. It was a natural economy, geared toward self-sufficiency. The nobility and clergy of the feudal system skilfully extracted every last ounce of service from the serfs. Indeed, given that this parasitic class represented around 10 percent of the population, it strongly shows that, even without major trade or organization, the feudal economy was far from primitive.

The feudal system spread technical advance throughout the lands. In contrast to classical times, where labor-saving devices were restricted to the city-states, in feudal times there was a more widespread use of iron, better plows, better mills, and better looms. But the localized nature of the economy prevented rapid progress. When feudalism spread over the untilled lands of Europe, the expansion overstretched itself, leading to an acute economic crisis from which it never really recovered.

Only in Europe, and especially between the eleventh and fourteenth centuries, the feudal system was more fully developed, complete with its own political and religious hierarchies, along with an associated knowledge and art.

It was an age of faiths. The system of feudal production kept the need for useful science to a minimum. It was not to rise anew until navigation, trade, and the growth of the towns produced new needs in the late Middle Ages. Meanwhile, intellectual effort was instead plowed into a radical new aspect of civilization: ordered religion.

The day of the organized clergy had dawned. The advent of the "peoples of the book" was a worldwide phenomenon. Not just for

Europe, but also for widely different regions, religion began to emerge as the prevailing social and political force. From the third through to the seventh century, we now see the growth of power and influence of Islam, Buddhism, and Christianity.

Once these new religions had peaked in terms of revolutionary fervor, they became essentially conservative, stabilizing organizations. Spiritual sleights of hand, the invoking of gods, myths, and prophecy, distracted the discontent from injustice and provided "celestial balance." The mission, whether implicitly or explicitly, was to control, to render the feudal social order palatable by showing how it was part of an unchangeable universe.

DOGMA AND SCIENCE

It is remarkable indeed that feudalism was such a stagnant system. Across the globe, the vacuum left by the collapse of the slave-owning plutocracy saw the rise of a feudal economy. It was an economy so fragmented and localized that it needed no radically new worldview, no paradigm was necessary to replace the ancient one that had served the classical world.

Plato and Aristotle prevailed. The human message of Jesus could not possibly continue to serve the purpose of the Church once it sought cultural domination. Plato was resurrected. His philosophy was already present in the message of St. Paul, whose influence on Christian thinking has arguably been more significant than any other New Testament author. And as the gospel of St. John shows, with its cult of the divine word, the logos, Platonism was key from the very beginning.

Science became stagnant. The victory of Christianity in the West meant that, from the fourth century on, intellectual life was articulated through Christian thought. Learning was confined to churchmen, and during the early Middle Ages, the history of thought over the lands of the disappearing Roman Empire was the history of Christian dogma.

The Church fathers had set about their mission impossible: to integrate the more innocuous elements of the ancient wisdom into Christianity. Much of the old philosophy had already found its way in, by stealth. But the Old Testament and classical culture were unhappy bedfellows. And as the Neoplatonists tried to crowbar in some safer aspects of philosophy, controversy was inevitable.

Heresy was born. Beginning in the fourth and fifth centuries, great disputes and heresies raged, particularly over Platonic ideas of the nature of the soul and its relation to corruptible bodies. Indeed, even the dispute of the nature of the Godhead itself was decided, as part of the great Arian heresy at the Council of Nicaea in 325 AD.

By the fifth century the deal was done. St. Augustine had brokered a settlement between philosophy and faith, and science was in the doldrums.

> When, then, the question is asked what we are to believe in regard to religion, it is not necessary to probe into the nature of things, as was done by those whom the Greeks call physici; nor need we be in alarm lest the Christian should be ignorant of the force and number of the elements: the motion, and order; and eclipses of the heavenly bodies; the form of the heavens; the species and the natures of animals, plants, stones, fountains, rivers, mountains; about chronology and distances; the signs of coming storms; and a thousand other things which those philosophers either have found out, or think they have found out ... It is enough for the Christian to believe that the only cause of all created things, whether heavenly or earthly, whether visible or invisible, is the goodness of the Creator, the one true God; and that nothing exists but Himself that does not derive its existence from Him.[6]

Rather simple, really. Augustine was very familiar with Greek science. He admired its accuracy and scope. But the pursuit of such scientific questions was incompatible with his faith. He delved no further. Augustine's less liberal contemporaries were more scornful of pagan science. And what follow-on fusion there was between such scriptural slavishness and Platonism helped forge an otherworldly faith, based on the conflict of good and evil, predestination, and ideas of hellfire and the Devil.

In contrast, the science of observation and experiment was far too worldly. On the one hand, it was hardly necessary for salvation. And on the other, through science's mere trust in the senses, it greatly devalued the worth of revelation. It was an attitude that was to endure for centuries. And in the days after the decaying Roman Empire, it was no time to turn over a new page.

Science was a victim of dogma. The absurdity of combining Classical thinking with the mythical Testaments, Old and New, was bound to produce fatalities. Scripture is at best self-contradictory, and was never meant as an analysis of nature. The sheer nonsense of trying to merge the immiscible, in the form of natural philosophy and scripture, was lethal to any lucid grasp of the universe. Faith and reason are irreconcilable. They cannot be squared without allegorizing the one or mangling the other. Not a recipe for honest thinking.

But the truth will eventually out. We should not thank the Church for saving the wisdom of the ancients. Science survived on its own merits, through its success in dealing with the real world. And it has endured, despite the attacks, despite the attempts to suppress it, despite the abortive cracks to unite it with conflicting beliefs.

As we will see with the controversy over Darwinism, progress with the theory of evolution was held back for decades, simply because the facts did not tally with the pages of Genesis. Undoubtedly, the average medieval churchman of his time did his utmost, according to his brief. But there were those who knew better.

Science crept through the Middle Ages. Its slow progress was not just due to the Church and the fetters of Christianity. It was primarily a question of economy. Under such feudal conditions, the Church's role was so defined, and the advance of science could have been no faster.

MEDIEVAL SCIENCE

During the medieval period came the decay, spread, recovery, and rebirth of classical Greek science. This is true not just for the West, but also largely for Asia, where science and technique were derived from the same worldview, embodied by Plato and Aristotle. For some seven hundred years, the civilized world salvaged the essence of Greek science and technique. So it was in India and China, Persia and Syria, Egypt and Europe.

In the Middle East came a cultural fusion. The Islamic world acted upon the impulse of Greek knowledge, forging it with their culture, and creating new beginnings in science. Islamic al-Andalus passed much of this on to Europe. Arabic astronomy was later adapted into the Copernican heliocentric model, for instance, and Al-Kindi's law of terrestrial gravity influenced Robert Hooke's law of celestial gravity, inspiring in turn Isaac Newton's law of universal gravitation.

There was a contrast in Christendom. What little science there was in medieval times was done by clerics. Islamic science was mostly practiced for utilitarian ends. European science, for the greater part, was carried out by priests, monks, and friars, for religious reasons. Even the greatest intellectuals of the day used their science merely to support revelation.

Inspired by the works of early Muslim scientists, the most famous Franciscan friar of his time, Roger Bacon, placed considerable emphasis on empiricism. Also known as Doctor Mirabilis, "wonderful teacher," Bacon was an English philosopher who is sometimes credited as one of the earliest European advocates of the modern scientific method.

Even towering intellects such as Bacon and the enigmatic Peter the Pilgrim, however, carried out their experiments in the name of the Church. They used method, rather than reason. They used experiment, rather than conjecture. But their work was largely demonstrative, rather than an attempt to truly investigate, let alone control, nature.

To be fair, they had little incentive and ample reason to be discouraged. Bacon spent a sizeable fortune on his researches. And despite the Pope's blessing, he was put in prison for his pains.[7] Having learned the lesson well, the inscrutable Peter the Pilgrim, though a pioneer in the study of magnetism, did not "care for speeches and battles of words, but pursues the works of wisdom and finds peace in them."[8] Pilgrim kept his head down.

The medieval contribution to science was meager, at best. Rather trifling contributions were made in fields such as natural history and minerals, optics, and sporting birds. There was even a semirespectable account of the rainbow, one that was not to be bettered until Newton himself. But the overwhelming impression was that of rather scanty spare-time work.

MEDIEVAL ASTRONOMY

In astronomy, it was the same story.

Aristotle was the last notable cosmologist of antiquity, and Ptolemy was its last able astronomer.[9] The work of these two thinkers cast the longest shadow over medieval astronomy and eclipsed all scientific thinking in the Middle Ages. Copernicus and Galileo might as well have been their immediate successors. For in the fourteen mostly feudalist centuries that separate the death of Ptolemy and the birth of Galileo, no enduring change was made to the body of astronomical works.

Ptolemy's magnum opus, *Almagest*, was the only surviving comprehensive ancient treatise on astronomy. Written in the second century AD, the treatise consisted of thirteen books, a state of the universe report on Aristotelian cosmology, and a guide to the complex motions of the stars and planetary paths. Hardly a "Rough Guide." Within its pages, Ptolemy claimed to marshal a model based on the observations of his predecessors, spanning more than eight hundred years. But astronomers suspected for centuries after that the models he created were divorced from observation and experiment.

Ptolemy had had a crack at solving the problem of the planets, and their insistence on tormenting the Greeks with their retrograde motion. His legacy was a rather farcical fairground of a universe, wheels within wheels, like one gigantic wind-up toy, complete with clockwork components. Aristotle had used nested spheres to explain away change in the cosmos, as dictated by the idealism of Plato. Ptolemy used epicycles.

As master mathematician, Ptolemy took up Plato's philosophical mission to excuse the evident irregularities of the paths of the planets across the ancient sky. Once more in pursuit of heavenly perfection, Ptolemy, like the geocentrists before him, had the planets rotate in circular motion at uniform speed. But his "solution" to the problem of the planets came in the form of a rather ridiculous, but cunning proposal: the notion that the planets were attached, not to Aristotle's concentric spheres themselves, but to circles attached to the concentric spheres, as shown in Figure 4.1, Schema 6.

These circles were called "epicycles," and the concentric spheres to which they were attached were known as the "deferents" (Schema 6a). The center of each epicycle traveled with uniform circular motion about the central Earth. The planets themselves orbited the center of the epicycle, again with uniform motion. The effect of this mildly bewildering idea is shown in Schema 6b. As the center of the epicycle

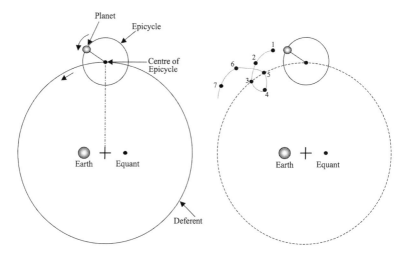

Figure 4.1 Schema 6: The Epicycle System of Ptolemy. In Schema 6a, the center of the epicycle moves counter-clockwise on the deferent. The planet moves counter-clockwise around the epicycle, whose motion is uniform with respect to the equant. In Schema 6b, the combined motion of the epicycle and planet is in the general direction indicated by the cycloid path, points 1 through 7. But, seen from the central Earth, the planet appears to move backward, the retrograde motion shown by points 3 through 5.

moves around the deferent, and the planet moves around the epicycle, the apparent position of the planet is that shown by the rather wavy path. Eureka! The retrograde motion of the planets is explained away by sleight of mathematics.

That an ancient astronomer with the talent of Ptolemy could convince himself that this elaborate scheme still amounted to uniform circular motion is testament to the influential curse of Plato. In particular, were Plato's three powerful ideas, completely wrong, but deeply ingrained into the mindset of ancient astronomers: that a stationary Earth sat at the squalid center of the universe; that heavenly objects were made from perfect material, unable to change their intrinsic properties, such as brightness; and that all motion in the heavens was uniform and circular.

However, retrograde motion was not the only outrage of the planets. As we know today, the planets move not in circles, but in ovals. So Ptolemy introduced a purely mathematical concept, the equant, to do away with any associated anomalies. As also shown in Schema 6 above, the equant point is placed so that it is directly opposite the Earth from the center of the deferent. The result: a planet was conceived to move with a uniform motion with respect to the equant.

There were drawbacks, however. With such a combination of abstract concepts, Ptolemy was able to account for the motions of heavens, especially the anomalistic paths of the planets. But even though his scheme succeeded, Ptolemy had violated the cosmology of Aristotle. The Earth was no longer the center of the universe. The paths of the planets and the rest of the cosmos were instead centered just off the Earth, in between it and the equant.

It was a fudge that few objected to. And that fact indicates a rather telling tale. Astronomy, in the hands of the idealists, had become an abstract sky geometry, divorced from observation and research. There was some considerable disquiet over another violation, however. The idea of the equant abused Plato's ideal of perfect circular motion, and this violation bothered thinkers a good deal more. Indeed, Copernicus was later to claim that his main aim was to rid the cosmological model of the monstrous construction of the equant.

By the time the Ptolemaic system was perfected, the ancient heavens required a clockwork of thirty-nine wheels with which to make it tick. When the wheel of the fixed stars was added, the count was raised to a full forty. Ptolemy's model, first catalogued in 150 AD, became the new standard bearer of ancient astronomy. When Christendom recovered ancient learning from the Arabs, it was naturally often in Arabic translation. Hence, Ptolemy's great tome is not

known by its Greek name at all, but by a shortened Arabic title it got
from a ninth century Moslem translator who called it *Almagest*, literally "The Great Book."[10]

Indeed, so little had changed in essence since Ptolemy's day, that
his system remained the dominant one acknowledged by academia in
Paradise Lost, English poet John Milton's seventeenth century epic
saga. The aim of his work was to "justify the ways of God to men."[11]
Milton assesses the Ptolemaic model with some ridicule:

> From man or angel the great Architect
> Did wisely to conceal, and not divulge,
> His secret to be scanned by them who ought
> Rather admire; or, if they list to try
> Conjecture, he his fabric of the Heavens
> Hath left to their disputes, perhaps to move
> His laughter at their quaint opinions wide
> Hereafter, when they come to model Heaven
> And calculate the stars, how they will wield
> The mighty frame, how build, unbuild, contrive
> To save appearances, how gird the sphere
> With centric and eccentric scribbled o'er,
> Cycle and epicycle, orb in orb.

Perhaps the ultimate word on Ptolemy's "quaint opinions wide" goes
to Alfonso X, the thirteenth century King of Castile. Also known by
his nicknames of "el Sabio," "the Wise," and "el Astrólogo," "the
Astronomer," Alfonso's view when first introduced to the Ptolemaic
system, was to heave a sigh, and say, "If the Lord Almighty had consulted me before embarking upon Creation, I should have recommended something simpler."[12]

Five centuries stood between Aristotle and Ptolemy. And yet when
their works became Latinized, those who read their thoughts were
spellbound by the coherence of their cosmologies. For men emerging
from the Dark Ages, the ancient wisdom was awesome in its scope,
dazzling in its brilliance. The first and overriding task was to assimi-
late the classical knowledge.

Ptolemy's grand system had neither successor nor rival. Both in the
Arab world and in Europe, the bulk of the technical astronomy during
the prevailing fourteen centuries was dedicated to the computation of
tables and the design of instruments, all in support of *Almagest*. Ptolemy
was transformed into almanacs, horoscopes, and planetaria.

To the contemporary eye, it is perhaps too easy to condemn medieval astronomers for their lack of critical thinking. Surely, it is obvious

to the careful and meticulous observer that the heavens revolve around the Sun and not the Earth? We might wonder why it was so hard to rid science of the geocentric mindset: the central Earth, the immutable heavens, and the uniform circular motion of the planets. After all, these obstacles led to dead ends, such as the abstract babble of Ptolemy's epicycles.

But it is not that straightforward. Consider the apparently simple conundrum: to prove heliocentrism, with the evidence available to Greek astronomers. Or, in other words, to prove the solar system was Sun-centered, even with the facts at hand in the pretelescopic medieval period. The situation had remained essentially unchanged, for the technology available to explore the sky had hardly advanced in that time.

To the naked-eye observer, the entire canopy of the heavens, stars, planets, and Milky Way, appears to rotate about a central Earth, rising in the East and setting in the West. The geocentric model explains simply this day-to-day procession. But even the more anomalous and the occasionally spectacular celestial events come under the ken of the Earth-centered model. Though the retrograde motion of the planets may seem problematic, lunar and solar eclipses may seem incredible, and episodic changes in a body's brightness may seem troubling, the canny gadget of Ptolemy's epicycle was just the slight adjustment needed to explain away the glitches.

As to other cosmic changes, such as the odd comet or extremely irregular supernova, these changes could also be excused. Aristotle had insisted that such phenomena were merely disturbances in the Earth's atmosphere, the sublunary sphere being the only realm in his cosmology subject to any change.

So from Plato on, hardly anyone doubted the geocentric model. If the abstractions did not precisely fit observation, reason and common experience condoned it. Philosophers and poets, encyclopedists and educators, noblemen and kings, all spoke of the universe much as Plato, Aristotle, and Ptolemy had described it.

Besides, there was the Church.

THE MEDIEVAL WORLD-PICTURE
AND THE THRONE OF GOD

The major trait of the assimilated Greco-Arabic medieval system was one of hierarchy. The cosmological scheme of Ptolemy and Aristotle became an unyielding, theological-physical world, a system of spheres and epicycles. The spheres of the Sun, Moon, and planets above were mirrored by a necessary counterweight, the underworld. Just as the Throne of God lay beyond the sphere of the fixed stars in heaven, the locus of Satan lay below the circles and pits of hell, deep in the bowels of the Earth itself.

The world was ordained as one of rank and place.

The tiered order of feudal society was reflected in the hierarchy of the medieval universe. Just as the great mass of serfs in feudal Europe had to put up with pope, bishops, and archbishops, the lowest ranks of mere angels had to contend with a celestial order. A cosmic hierarchy of nine choirs of angels, in fact: seraphim, cherubim, and thrones; dominations, virtues, and powers; principalities, archangels, and angels. It is enough to make your head swim.

There was a heavenly order and a social order, a place for everything and everyone knew their place. Each rank of angel had a job in helping run the universe, each contribution aiding the motion of the planetary spheres. The lowest order of angels belonged to the sublunary sphere, and naturally had most to do with the human beings beneath them. The four elements also obeyed a divine order. Earth below, water above, air higher still, and fire, the noblest element, the highest.

The entire medieval worldview was a compromise. The Aristotelian picture of a permanent cosmos on the one hand, and the Judeo-Christian portrayal of a divinely created world on the other. It was an interim world. One that could be destroyed by a single act of God, and that played to divine rules—a stage upon which a man's life is played out, salvation beyond the stars, or damnation to the pits of Hell on Earth, at the squalid center of the universe.

This complex cosmos was simultaneously ideal and rational. It was a fusion of the certain truths of scripture, with the logically founded inferences of the ancient wisdom. Though scholars differed on this point or that, the true picture of the world had been solved for all time. It seemed possible, after all, to have a universe that was at once realistic, reasonable, and theologically robust.

And it was nonsense.

Shedding light on this medieval worldview is crucial. For it illuminates the way in which an assault on any integral part of the picture becomes something far more serious. Such an attack is not a mere intellectual criticism. It morphs into a potential assault on the entire order of society, religion, and the universe itself. And given such potential it is reasonable to expect a robust defense, from the massed power of the entire Church and State. Just as Galileo was to discover.

THE TRANSFORMATION OF THE MEDIEVAL ECONOMY

Slowly but surely, Europe became distinct. As the assimilation of the Greeks by the Islamic world was passed on, science and technique progressed against an increasingly unstable feudal economy by the late Middle Ages. Consequently, an emerging European science began to challenge the cramping straightjacket of Christianity, the dogmatic religion that had survived the breakdown of the old world.

The medieval hierarchical system of thought was, of course, necessarily conservative. Left to its own devices, it would no doubt have been conserved to this day. But it was not left to itself. Much as the medieval system of thought had a great tendency to stasis, medieval trade could not stand motionless.

Once more, the economy was the driver.

The productive forces of trade play the most important part in social change. The feudal system held the seeds of its own demise and disintegration. Improvements in technique and transport transformed the economy, from one based on approved service to one based on commodities and money. And the technical aspects of this economic revolution were the driving force in the creation of a new, radical, and experimental science, replacing the stagnant, idealist science of the Middle Ages.

Renaissance thinkers were soon to face brand new challenges, problems the ancient wisdom was incapable of dealing with. So the conscious human intervention came later, when the changes in technique and economy of the later medieval world led to the radical rise of capitalism. But a number of vital technical advances began during the Middle Ages, and indeed represent a significant foundation for the scientific revolution.

Medieval society was static and hierarchical. Nonetheless, technical change did occur. That it remained unnoticed for so long is testament to the domination of clerical chroniclers, any innovation being lost in the ecumenical signal-to-noise ratio. They were not lost to the merchants, however. One example of such significant groundwork is the logbook of Villard de Honnecourt, a thirteenth century master mason, that contains the commerce and sketches of various gadgets and devices.[13]

Precious few medieval scholars paid much attention to scientific concerns, fewer yet attempted to comprehend technical matters. However, one exceptional scholar was Roger Bacon, whose acclamation of Peter the Pilgrim speaks volumes for the dormant thread of materialism running through these otherwise superstitious days.

He knows natural science by experiment, and medicaments and alchemy and all things in the heavens or beneath them, and he would be ashamed if any layman, or old woman or rustic, or soldier should know anything about the soil that he was ignorant of. Whence he is conversant with the casting of metals and the working of gold, silver, and other metals and all minerals; he knows all about soldiering and arms and hunting; he has examined agriculture and land surveying and farming; he has further considered old wives' magic and fortune-telling and the charms of them and of all magicians, and the tricks and illusions of jugglers. But as honour and reward would hinder him from the greatness of his experimental work he scorns them.[14]

New inventions set in train a revolution in technique. The vast importance to the entire world of Chinese technical innovation is now well recognized. Many inventions showing up in the tenth century or later in western Europe were fully developed in China by the early centuries of this period.

Paper and printing, gunpowder and cannon, horse harness and sternpost rudder, clock and watch, watermill and windmill—all brought tremendous improvements in productivity. And this revolution aided the collapse of the feudal economy through increased productivity and trade. Better farming meant more surplus with which to trade. Better transport led to a more discerning use of land. Trade in turn enhanced the towns and the status of the merchant class.

Over the lands of Europe far and wide, the cumulative influence of improved trade was felt. Enhanced productivity and transport increased the surplus of the village and the amount of produce that could be consumed there. Wealthy peasants and urban merchants became stronger. They represented a growing market, stimulating the manufacture of goods. The Middle Ages had reached a turning point.

NAVIGATION, PIRACY, AND PLUNDER

Two Chinese navigational inventions, the compass and the sternpost rudder, were to have a global effect at sea. Long voyages became more viable. The seas were thrown open. Open to exploration. Open to a colossal expansion in trade. Open to piracy. And open to war.

Scientifically, the consequences of better navigation were profound. The oceans open to navigation meant a need for a more accurate astronomy. Better observations, better charts, and better instruments. Open-sea navigation also raised the need for a more predictive astronomy, a brand new quantitative geography, and the desire for devices that could be used onboard ship as well as on land.

The obsession with longitude began. Deep-sea navigation also raised the urgent need to know where on Earth they were. Mariners and explorers for most of history had struggled to determine precise longitude. Latitude was no problem; observing or predicting the positions of the Sun, Moon, planets, and stars could easily calculate it. Instruments such as the quadrant or astrolabe were tools of the mariner's trade in finding latitude. They read the inclination of the Sun or of charted stars. Longitude presented no such manifest means of study.

All the great astronomers of the day were to try their hand at determining longitude. The need for compasses and other devices for navigation ignited a new skilled industry, especially that of the card and dial makers.[15] The ensuing influence on science was huge. It set new standards for precision, and many prominent scientists were also instrument makers, including Newton, Watt, and, of course, Galileo.

The great sea voyages started with the Portuguese explorers around 1415 and opened the planet to European capitalist enterprise. The voyages were the fruit of the first deliberate use of astronomy and geography in the pay of glory and profit, with a practical eye for sugar plantations, slaves, and gold. German scholars such as Peurbach and his pupil Regiomontanus of Nurnberg, later assisted by Albrecht Durer, refined the application of astronomy to navigation, and initiated a drive for astronomical tables plain and precise enough to be of immediate use to the ocean navigators.

Theory and practice met at the court of Prince Henry the Navigator at Sagres, where Moorish, German, and Italian experts planned new voyages with hardened Portuguese and Spanish sea captains. The Turkish stranglehold on eastern trade raised the compelling idea of venturing into the Indian Ocean by some way other than the Red Sea. Strategists debated about two promising alternative routes. The first, to round Africa, was profitably realized by the Portuguese in 1488, though India was not accomplished by Vasco da Gama until 1497.

The alternative course, volunteered by the astronomers and theoretical geographers such as the Florentine Toscanelli was to head west over the uncharted ocean in search of China at the other side of the world. Now to reason such a hypothesis is one thing, but to sail straight out to sea quite another. In the popular imagination a myriad of potential fates might befall such adventurers; they might sail on without end; or, they might plunge off the edge of the world. Not a single soul anticipated there might be an entire continent in the way.

The one man fit for such a gamble has often been regarded the Sovereign Navigator, and the most opportune adventurer. Christopher

Columbus was many light-years from being a scientist, but he did have a brilliant vision—that he may well, by seafaring west, unearth new worlds, or even discover "a new heaven and a new earth." And it was this dream, part spiritual, part scientific, that enabled Columbus to finally seduce the venture capitalists.

The distinction between the successive voyages of the Portuguese round Africa and of Columbus staking all to set sail directly across the Atlantic is emblematic of that between technical improvement, relying on a gradual advance of tradition, and a scientific one, which rationalizes to provide a revolutionary break with ritual. Columbus's mystical motives notwithstanding, the finance he received for his voyage was invested on the potency of a practical appraisal of the booty anticipated from the proof of a scientific proposition.

The simultaneous unearthing of the New World and the aged and prosperous worlds of Asia, with all their exotic goods and customs, rendered the classical world provincial.[16] Adventurers were won over by the knowledge of accomplishments that the ancients had never imagined. The architects of the Renaissance wished for a new age, and by the middle of the sixteenth century they could believe they had attained it. The rejuvenated discipline of navigation, now unlocked to study and explanation, needed new methods of exploration.

The movement of the heavens now had cash value.[17]

CHAPTER 5

THE GREAT CHAIN OF BEING

HARD LIMITS IN SPACE AND TIME

Just as the typical medieval town was walled in, so was the medieval universe: a walled-in cosmos, bounded between Heaven and Earth, closed to the ravages of change and time. Continuity was key. It was a connected cosmos, emanating from the realm of the Godhead, through the nested concentric planetary spheres of crystalline perfection, and down to the dark and lowly corruptible Earth at its center.[1]

The simple two-tier system of Aristotle had become the ornate pageant of divine creation that was The Great Chain of Being. The easy split of the cosmos into sublunary and supralunary regions had remained. But within its broad boundaries now sat the cornucopia of creation, an infinite procession of links, stretching from God down to the lowliest form of life.

The contrast of permanence and change in Aristotle's original system had also been preserved. This scale of being, or *scala naturae*, was a strict hierarchical system, ranging from the highest perfection of the unchanging Spirit, who sat at the top of the chain, down to the fallibility of flesh at the core, mutable and corruptible Man.

It was a natural order, of sorts. Elements that sat at the foot of the chain, such as rock from which the Earth is made, merely existed. Moving up the chain, each successive link enjoyed more positive attributes than those below. Plants, then, possessed life, as well as existence; animals enjoyed the additional qualities of motion and appetite.

Every imaginable creature and object had a place in this great scheme of things. But each position was determined in a rather anthropocentric way, often according to its utility to man. Wild beasts were superior to domestic ones, since they resisted training. Useful creatures, such as horses and dogs, were better than docile ones, such as sheep. Easily taught birds of prey were superior to lowlier birds, such as pigeons. Edible fish were higher up the totem pole than more dubious and inedible sea creatures.

Even aesthetics came into it. Attractive creatures such as dragon-flies and ladybirds were considered more worthy of God's glory than unpleasant insects such as flies and, no doubt, dung beetles. The poor snake languished at the very bottom of the animal segment, relegated there as punishment for the serpent's actions in the Garden of Eden. Some aspects of the Chain persist in popular culture today: the lion is still considered the king of all creatures, the oak the king of all plants.

It was not just unthinkable to abandon your place in the great continuous Chain; it was impossible.

This strict sense of permanence was crucial to the conception of the world. And it was easy to see how the system became a useful means of control in feudal society. The Great Chain was used to justify the doctrine of the Divine Right of Kings, the idea that the monarch was subject to no earthly authority. The Chain had its reflection in this authoritarian social order that saw kings at the pinnacle, aristocratic lords below, and the great mass of peasants way down the societal scale. The king derived his right to rule directly from above, and was not subject to the will of his people, the aristocracy, or any other estate of the realm, but to the will of God.

Spiritually, man was a special instance in the Chain. He was both mortal in flesh and potentially pure in spirit. The resolution of this struggle between his two aspects meant that he could either go the higher noble way of the spirit, or be dragged down to the devil, the way of all flesh, just as Lucifer himself had fallen.

The conception of the *scala naturae* is summed up in the following quote, referenced often throughout the Middle Ages, and written by fifth century philosopher Macrobius, one of the last pagan writers of Ancient Rome:

> Since, from the Supreme God Mind arises, and from Mind, Soul, and since this in turn creates all subsequent things and fills them all with life … and since all things follow in continuous succession, degenerat-ing in sequence to the very bottom of the series, the attentive observer will discover a connection of parts, from the Supreme God down to the last dregs of things, mutually linked together and without a break. And this is Homer's golden chain, which God, he says, bade hang down from Heaven to Earth.[2]

The Great Chain of Being was, in part, a theory of biology, a theory of the generation of "sentient and vegetative creatures." And as such it was a vastly influential concept, until it was eclipsed by the theory of evolution, many centuries after. As a result, we shall meet Macrobius's idea of a Chain "linked together and without a break" in more detail later, since the concept lay at the center of the nineteenth century scandal of extinction.

Once more, the influence of Plato and Aristotle reigns supreme. Plato's principle of plenitude dominates the upper part of the Chain. This idea of Plato's taught that the universe contained all possible forms of existence. Everything that can exist, does exist. For a universe in which some possible species of existence was not realized would be imperfect, since it would not be full. This Platonic inspiration is key to the idea that the Most Perfect Being would "emanate" and "overflow" into the world, creating forms and copies of its true self, replicating down in a descending series, to the "last dregs of things."

And if the upper part of the Chain was Platonic, the lower part evinced the influence of Aristotle's biology, revived around 1200 AD, with its ideas of continuity and gradation.

> Nature passes so gradually from the inanimate to the animate that their continuity renders the boundary between them indistinguishable; and there is a middle kind that belongs to both orders. For plants come immediately after inanimate things; and plants differ from one another in the degree in which they appear to participate in life. For the class taken as a whole seems, in comparison with other bodies, to be clearly animate; but compared with animate to be inanimate. And the transition from plants to animals is continuous; for one might question whether some marine forms are animals or plants, since many of them are attached to the rock and perish if they are separated from it.[3]

ARISTOTLE'S BIOLOGY

Aristotle was the first great encyclopedist. His aim was to provide an account of every aspect of nature, all zoological forms from zoophytes to man, in an ordered taxonomy of the natural world. His "principle of continuity" allowed him to arrange all living things into a hierarchy, reflecting degrees of perfection. It also made it possible to link the two halves of creation—the sublunary and the celestial—into a single continuous scale of being.

Rather than being a moralist, like Plato and Socrates, Aristotle had been a scientist, and a logician. Instead of striving for a type of social utopia in an independent aristocratic republic, as did the moralists,

Aristotle was a pragmatist, content to make the best of things as they were. He was the philosopher of common sense, his Chain an order of commonplace.

Ever the pragmatist, Aristotle even refused to discuss the ultimate origins of the natural world. The world always was as it is, since that is the logical way for it to be. Of course, such a steady-state view of the universe was to prove rather problematic later, when it was adopted by the Catholic Church as the philosophical basis for its cosmology. So they simply stuck a sudden creation event at the beginning and a sudden destruction at the end.

Nevertheless, Aristotle had set about his taxonomy of the universe as a logician. His classification of the physical world was in the image of an ideal social world, where subordination was the natural state. His code of logic was to sort things according to resemblance and difference. His approach was to ask questions such as, "What is a thing like?"—its genus, and, "How does it differ from other things that are like it?"—its differentia.

The key to understanding the world, suggested Aristotle, was physics. But his was a very *biological* physics. Rather than being about lifeless matter in motion, Aristotle's physics was an interpretation of the world as if everything were actually alive. His explanation of all things, from why stones fall to why some men are slaves, boils down to the same cause, "it is their nature to do so." Aristotle's thoughts on the nature of things, and their structure, became the standard ancient system of biology, handed down through the centuries.

The subsequent story is, in many ways, the tale of Aristotle's overthrow. His comprehensive insight on the *nature* of everything, translated very easily into "the will of God." In essence, it is the same explanation, only less scientific. This divine mantra was woven into the fabric of the cosmology that Galileo and Darwin helped depose.

And it was fundamental to the Great Chain of Being. The "will of God" enabled creation by a process of degeneration by descent, the very opposite of the idea of evolution that was to be developed by the likes of Darwin. Nonetheless, the insufficiency of Aristotle's biology is somewhat saved by the breadth and quality of his observations. The guiding light of his biology is in the notion that everything in nature reaches up to achieve what perfection it can, by degrees. It is this notion that drove Aristotle to set down a scale of nature, with minerals at the bottom, then plants, then more and more perfect animals, with man at the pinnacle.

In some senses, the Chain of Being may be thought of as implying an evolution of sorts, if Aristotle was to be taken literally in believing nature strives for perfection. But in fact he was sure that nothing really changed in the natural world. Species were eternal and permanent signposts on the road to perfection, or imperfection. Indeed, imperfection was Aristotle's tendency. He was more likely to see an animal as an imperfect man, and a fish as an imperfect animal, than vice versa.

For two thousand years the book of life was slammed shut.

The world of Aristotle was one in which everything knew its place. In his physics, motion only occurred when an object was "out of place" and was returning to its proper position. So stones fall to the ground as sparks fly skyward to join heavenly fires. In his biology, Aristotle held the idea of final causes. It was in the nature of birds to fly, and fish to swim. That is what such animals *are for*. If other, material and effective causes could be identified, he thought them inferior to final causes.

This doctrine also proved a curse to science. For it engenders the notion that nature is somehow explained by claiming an end to each means, an ultimate goal to each phenomenon, without having to get to grips with the reason why each phenomenon actually occurs and works. In biology, rather than physics, at least the idea of final causes was in some ways workable. For it is an expression of adaptation of organisms to their environment. But even in biology the effect was finally befuddling. All one need do is guess at an organism's purpose, and hang the rest of it.

And yet the latent notion of evolution was present in Aristotle's work. It was encapsulated in his idea of *potentiality*. Even the crudest matter was capable of any form, *potentially*. And these forms represent a purpose of perfection, which may not always be attained. To Aristotle, each substance also had an essence. So, substantially, a man has two legs, but they do not define his essence. For he could lose a leg and still be a man.

The notions of *essence* and *potentiality* are biological in quality. They articulate the lower and upper limits of what each member of a species can achieve. In its essence, it merely manages to exist, when it reaches its potential, of course, it is exhibiting the full range of its potential powers. In this way, Aristotle opens the way to the idea of evolution of forms, from the imperfect to the perfect.

But evolution was an idea yet to be born. The dominant ancient authority of Aristotle, allied to that of the book of Genesis, held back any early ideas of evolution, such as those of the Atomists, for more than two millennia. Meanwhile, the notion of differing grades of perfection also had a social use. It justified the belief that some men were made to be masters, while others were naturally slaves. And if natural-born slaves were too stupid to realize their birthright, wars to enslave them were naturally just.

FROM GOD'S THRONE TO THE MEANEST WORM

In the Middle Ages, the Great Chain of Being was molded into a more expressly Christian shape.

For heaven and for Earth, scholars put the Chain to divine rights. Pseudo-Dionysius, the anonymous turn of the sixth century Neoplatonist philosopher, set about fixing the order of angels in the upper reaches of the Chain who kept the heavens in motion. The Seraphim turned the Primum Mobile, the Cherubim the sphere of the fixed stars, and so on, down to the lower angels, who had the misfortune of looking after the Moon.

Meanwhile, St Thomas Aquinas, the immensely influential thirteenth century philosopher, developed the connectivity of the Chain. The key, said Aquinas, was in the dual nature of man. In the continuity of all things, the lowest member of the higher genus is always found to border upon the higher member of the lower genus. This principle was as true for zoophytes, which are half plant, half animal, as it was for man, who

> has in equal degree the characters of both classes, since he attains to the lowest member of the class above bodies, namely, the human soul, which is at the bottom of the series of intellectual beings—and is said, therefore, to be the horizon and boundary line of things corporeal and incorporeal.[4]

Aristotle had never exactly defined the criteria with which the degrees of perfection were determined. But medieval scholars were gritty in their resolve to extend the Chain, drilling through the four elements, down into the dirt of inanimate matter. Where no obvious ciphers could be found to decide an object's difference, the black arts of astrology and alchemy showed "correspondences" and "influences." So, each planet became connected with a color, a metal, a day of the week, a stone, and a plant, each association defining their place in the Chain.

Lower still lurked Lucifer. The further descending extension of the Chain led into the supposed conic cavities that skulked in the bowels of the Earth. Here chains of devils occupied the nine circular slopes, which replicated the nine heavenly spheres. Lucifer himself loitered at the apex of this downward cone, the exact center of the Earth, and the sorry end of the Chain.

It was a system of the world that was not so much geocentric as it was "diabolocentric."[5] A system with hell at its heart. For although the Chain was continuous in nature, and although the heavens were incorruptible, the Earth sat among "the filth and mire of the world, the worst, lowest, most lifeless part of the universe, the bottom storey of the house."[6]

The potency and power of the Chain as a vision of animate creation, and man's place in it, is testified in time. It held as great a grip on the imagination of Dante as it did the Elizabethan poets' three centuries later. And two hundred years beyond that, it still managed to move Alexander Pope, the greatest English poet of the eighteenth century, to write,

> Vast chain of being! which from God began,
> Nature's aethereal, human, angel, man,
> Beast, bird, fish, insect …
> from Infinite to thee,
> From thee to nothing.—On superior pow'rs
> Were we to press, inferior might on ours;
> Or in the full creation leave a void,
> Where, one step broken, the great scale's destroy'd;
> From Nature's chain whatever link you strike,
> Tenth, or ten thousandth, breaks the chain alike[7]

Pope's warning is clear. The cost of a break in the chain, no matter where that break were made, would be the collapse of the entire cosmic order. Here again, as with medieval astronomy, we can now comprehend the common conception of the medieval universe. It is a static world, impervious to change. A cosmic inventory of creation, cast in permanence by God.

Crucially, man is at the midpoint in creation. And just as any cosmic shift in the position of the Earth would imply moving the Throne of God, any change in the Chain of creation, however small, would have cataclysmic consequences. For the rigid, graded hierarchy of life would come crashing down. We can already begin to see the havoc extinction would wreak on such a system.

In biology and society, there was no room for change. No evolution of species, and no social progress. A person would seek mobility in this life only in vain. For movement up or down the Chain is granted only after death. In this world, one's destined rank and place is final.

And yet the system was rife with inconsistency. The lowly material world was supposedly one of mutability. But the Chain brought blessed immutability to Earth, with an inert hierarchy as applicable to human affairs as it was to animal, vegetable and mineral. As Elizabethan poet and explorer Walter Raleigh put it:

> Shall we therefore value honour and riches at nothing and neglect them as unnecessary and vain? Certainly not. For that infinite wisdom of God, which hath distinguished his angels by degrees, which hath given greater and less light and beauty to heavenly bodies, which hath made differences between beasts and birds, created the eagle and the fly, the cedar and the shrub, and among stones given the fairest tincture to the ruby and the quickest light to the diamond, hath also ordained kings, dukes or leaders of the people, magistrates, judges, and other degrees among men.[8]

Indeed, the Chain was reflected in the rule of law that had a stranglehold on Elizabethan society, from top to bottom. Atop the chain of command was the monarch, who enjoyed rather draconian powers of punishment. Next came nobles, high clerics and gentlemen, in that exact order. Further down were the bourgeoisie, that fast rising class of wealthier merchants who would soon help transform the entire system. Then came small farmers, known as yeomen, and finally artisans and common laborers.

This social system was so exacting in its measures that it dictated what each person was allowed to eat and wear. The amount you were allowed to eat depended on status. A cardinal was permitted nine dishes at mealtime, while those earning less than £40 a year, the great majority of people, were allowed only two courses, and soup. Anyone found scoffing meat at Lent could be sent to jail for three months.

A measure of great legal restraint was also used to dictate who could wear what. These so-called sumptuary laws allowed a person with an income of £20 a year to dress in a satin doublet, but not a satin gown. Meanwhile, someone lucky enough to be bringing in an income of a hundred pounds a year could wear all the satin they wished. The number of other restrictions on the amount of fabric one could wear, and the ways of wearing them, were almost beyond measure.

So, by the late Middle Ages there was a greater horror of change even than in the ancient Greek world of Plato and Aristotle. It was a time when people were only too well aware of nature's wrath, red in tooth and claw. The darkest scourge was the plague, which stalked Europe from the fourteenth century, returning every generation with varying virulence and mortalities until the 1700s.

And if the plague didn't get you, the embattled citizens of Europe ran the risk of tuberculosis, rickets, measles, smallpox, dysentery, scrofula, and a host of fluxes and fevers. Unlike the notional Great

Figure 5.1 Der Doktor Schnabel von Rom, Doctor Beak of Rome, an engraving by Paul Fürst, 1656.

Note: The beak is a crude gas mask, crammed with substances (such as spices and herbs), and donned by physicians in a vain attempt to ward off plague and pestilence. The costume came complete with a waxed long cloak, eyes secured behind crystal, and a rod with which to point at what should be attended to. If the plague didn't kill you, the mere sight of the doctor surely would.

Chain of Being, these rank maladies were no respecters of position or place. It was a literally dreadful age.

Renewed religious fervor and fanaticism bloomed in the wake of the Black Death. And in the centuries of the late Middle Ages, the uncertainty of daily survival created a general mood of morbidity. It was easy enough to believe in the Earth's lowly position in the Chain, given "the filth and mire of the world." The catastrophic conditions of the age, its mood of despair, were hardly conducive to a rational, inclusive, evolutionary view of the universe, and man's role in it. But this petrified vision of the world, with its pedantically layered structure—rigid, static, and shrink-wrapped against the horrors of the "Black Death of Change"—was soon to crack down to its foundations.

GREEK GROUNDHOG DAY

The ancient Greek notion of time left little room for the concept of evolution. Philosophers such as Plato, Aristotle, and even Pythagoras, believed that time was cyclical, and that the history of the cosmos was made up of a series of "great years," as they called them. Each cycle, of unnamed length, ended in a planetary conjunction, which unleashed an apocalypse. Then, a new cycle began, out of the ashes of the old.

Time was a kind of cosmic Groundhog Day.[9] Essentially, the past was as closed and confining as the two-tier universe in which Aristotle kept cosmic space captive. It was part and parcel of a general feeling that history was eternal. There was no real beginning in time, since the "great years" ran endlessly on. As the Aristotelian, Eudemus of Rhodes, put it to his pupils, "If you believe the Pythagoreans, everything will eventually return in the self-same numerical order, and I shall converse with you staff in hand, and you will sit as you are sitting now, and so it will be in everything else."[10] A sobering thought.

It's not as if the ancients were entirely ignorant of geology. They knew that, within their great years, the world was subject to gradual change. An island dweller, such as Pythagoras, with the waves at his feet and the mountains at his back, would have been well aware of the attrition that water and time brought to the beaches. As early as the sixth century BC, thinkers such as Thales of Miletus, one of the Seven Sages of Greece,[11] knew that mountains were thrown up from what once was a seabed, and that in turn the hills could be worn down by wind and water. Both insights are foundation stones of modern geology.

The Greeks were also among the first fossil hunters. Another philosopher of the sixth century BC, Xenophanes of Colophon, figured that fossil seashells showed land was once under water. Indeed, many Greek thinkers were familiar with the peculiar fact that fossils of sea creatures were to be found on hilltops, far from the sea. But Aristotle proposed that fossil fish were merely the remains of fish that had become marooned and died while foraging for food in underground caverns.

For Aristotle, the universe was permanent. And the dominant ancient Greek doctrine of cyclical time had a rather pernicious effect on the idea of an evolving nature. After all, why try engaging with the genuine extent of natural history if it was merely made up of an eternal series of cycles, punctuated by catastrophes that may well have destroyed all evidence? Surely it would be impossible to find the true age of the Earth and uncover its evolving anatomy?

But the idea of a closed and circular time met its nemesis in Christian dogma. The eternal sweep of Greek time was shrunk down into a bite-size history, one that was linear rather than cyclical, one with a beginning in Genesis and an ending in Revelation. Scripture provided a measure of time, finite in scope and reckonable to those who should wish to do so.

Church scholars began to add up the "begats," the long procession of scriptural births and deaths, found in the Christian Bible. The fashion for doing so seems to have started with the father of Church history, Eusebius, chairman of the Council of Nicaea in 325 AD. It was Eusebius who claimed that 3184 years had elapsed between Adam and Abraham. Johannes Kepler caught the bug, too. The great German astronomer had estimated the date of Creation at about 3993 BC. Isaac Newton followed suit. He put the date at 3998 BC.

This specious technique was raised to an art form by the seventeenth century bishop of Armagh, James Ussher. This primate of Ireland famously concluded in 1658 that "the beginning of time, according to our chronologie … fell upon the entrance of night preceding the 23rd day of October, in the year of the Julian calendar, 710,"[12] or 4004 BC in modern terms. There are no doubt still some who believe such twaddle to be true. But the age-dating techniques of these Christian chronologists did have some lasting worth, it seems. They unconsciously paved the way for further, more scientific, enquiries about the genuine extent of the past.

The Fossil Record and the Terror of Time

There were far more important factors than Christian dogma, however—economy and industry.

By the end of the eighteenth century, science had begun to secure its dominion over nature. Newton's system of the world established its authority in the clanging new workshops of the world. The *philosophical* engine, the early steam engine, drove locomotives along their metal tracks; the first steamships crossed the Atlantic; the great transport magnates were building bridges and roads; telegraphs ticked intelligence from station to station; cotton works glowed by gas; and a clamorous arc of iron foundries and coal mines powered the Industrial Revolution.

Great engines of change were turning over the soil of the world. The speculation about the Earth and its fossils grew steadily in the eighteenth century, along with a great fascination with nature in general. And the emergent sciences of geology and biology did much to locate humanity's place in the depths of time, as astronomy did to chart our position in cosmic space.

The word "biology" entered the lexicon of science. This "science of living bodies" was quite distinct from the previous practice of natural history, with its focus on the three kingdoms of nature: animal, vegetable, and mineral. The Chain of Being had been part of that natural history tradition. The new word "biology" bore witness that *discontinuity* had replaced the earlier idea of the smooth contours of the Chain, where the most primitive organisms gradually faded into fossils, gems, and minerals.

The steam engine opened up the veins of the Earth, as the builders of the Great Wall of China had once put it. And in the developing industrial nations such as England and Germany, it was possible to dig deeper than ever before. Nascent field geologists helped manage the excavation of strata, laid down over hundreds of millions of years of Earth history, as yet an untold story.

The geologists soon learned how to read the rocks, however, as the French naturalist George Louis Leclerc explained in 1778:

> Just as in civil history we consult warrants, study medallions, and decipher ancient inscriptions, in order to determine the epochs of the human revolutions and fix the dates of moral events, so in natural history one must dig through the archives of the world, extract ancient relics from the bowels of the earth, [and] gather together their fragments ... This is the only way of fixing certain points in the immensity of space, and of placing a number of milestones on the eternal path of time.[13]

The geologists decoded nature's cipher, the language in the stones, and began amassing the evidence of the long history of planet Earth. The English geologist William Smith, consulting engineer for the Somersetshire Coal Canal in 1793, noted that "the same strata were found always in the same order and contained the same fossils,"[14] and the geologists realized that the world's natural history could be gleaned from the fossil sequence contained within the rocks.

But there was trouble ahead. The fossil record soon began churning out signatures of beasts no longer found on Earth, challenging biblically literal accounts of natural history. These beasts seemed to have no living counterparts, which produced a problem for those who derived their Earth history from Genesis, with the belief that all of animate creation was born at the same time and that none had become extinct.

At first, the growing evidence in the fossil record would not sway some. Wait awhile, they suggested. Soon enough, living specimens of all assumed-dead species would crop up, maybe in far off lands to which they had roamed in the years since the strata were formed. Thomas Jefferson, for example, urged pioneers heading west to search for the woolly mammoth. A deluded evangelical naturalist even reported having heard one trumpeting through the dark forests of Virginia.

And yet the death roll of extinction grew. The French zoologist George Cuvier, who helped found the science of paleontology, had by 1801 identified twenty-three extinct species in the fossil record. The word "extinction" found a place in the lexicon of science, too, ringing out in churches and chapels, far and wide. Today we understand that 99 percent of all species that have lived on planet Earth have since perished.

The new geology revolutionized the world. Its effect spread far beyond the scientific horizon, destroying established truths, forcing everyone to confront the terrible extent of history. Biblical literalists sought refuge in The Great Chain of Being. But the Chain was no more robust than its weakest link. Indeed, the very wholeness of the Chain was proof of the glory of God. There could be no "missing link," a term later appropriated by the evolutionists themselves.

As English philosopher John Locke wrote:

> In all the visible corporeal world we see no chasms or gaps. All quite down from us the descent is by easy steps, and a continued series that in each remove differ very little one from the other. There are fishes that have wings and are not strangers to the airy region, and there are some birds that are inhabitants of the water, whose blood is as cold as fishes … When we consider the infinite power and wisdom of the Maker, we have reason to think that it is suitable to the magnificent harmony of the universe, and the great design and infinite goodness of the architect, that the species of creatures should also, by gentle degrees, ascend upwards from us towards his infinite perfection, as we see they gradually descend from us downwards.[15]

The passage from Locke illustrates clearly the damage to be done by any notion of extinction in the minds of the pious. "It is contrary to the common course of providence to suffer any of his creatures to be annihilated,"[16] as the Quaker and Royal Society botanist Peter Collinson maintained. Likewise, the seventeenth century English naturalist John Ray feared that evidence of "the destruction of any one species," would result in "a dismembering of the Universe, and rendering it imperfect."[17]

So whither the Great Chain of Being? The idea of extinction and their associated "missing links" were to prove fatal. Like meteors, comets, and sunspots in astronomy, the "missing links" were monstrous deviations, fell portents of undesired change, and proof that something is very wrong in a system that is supposedly framed by the hand of God.

In December 1831, a young naturalist set sail on *The Beagle*. Charles Darwin would soon become a chance locus not only for the dissolution of the Chain, but also for a revolution in science which struck at the heart of humanity itself.

PART III

THE REVOLUTIONS: THE WEAPONS OF DISCOVERY

CHAPTER 6

THE TELESCOPE AND GALILEO

Lying on the torrent-like Arno river, in the hilly wine-growing Italian region of Tuscany, is the beautiful and historic city of Florence. Originally established by Julius Caesar in 59 BC, Florence is renowned for its pivotal importance in the Middle Ages and the Renaissance, especially with regard to architecture and art. A center of medieval trade and commerce, Florence is often considered the birthplace of the Italian Renaissance.

One of the many jewels in the crown of this "Athens of the Middle Ages" is the principal Franciscan church of Florence, the Basilica di Santa Croce, or Basilica of the Holy Cross. Situated on the Piazza di Santa Croce, almost one kilometer south east of the Duomo with its landmark dome,[1] the Basilica di Santa Croce is the burial place of some of the most illustrious Italians.

Renaissance painters Michelangelo and Donatello lie here, as do twentieth century Nobel Prize winning scientists Guglielmo Marconi and Enrico Fermi. And there are the funerary monuments of Dante and Eugenio Barsanti, coinventor of the internal combustion engine. Little wonder the basilica is known also as Pantheon dell'Itale Glorie, the Pantheon of the Italian Glories.

Perhaps the most curious pairing is that of the Renaissance political philosopher Niccolò Machiavelli and the one Tuscan whose legend loomed large in the history of the scientific revolution: the physicist, astronomer, and mathematician, Galileo Galilei.

So much is attributed to the great Galileo, and yet so little is supported by historic fact. Claims made for science of his outstanding "genius" rely mostly on inventions he never made, theories he never instigated. He did not invent the pendulum clock. Neither did he invent the microscope, nor the thermometer. He did not discover the law of inertia, nor did he toss two cannonballs from the leaning tower of Pisa.

Galileo did not invent the telescope. Nor did he make any great contribution to astronomy. He did not discover the sunspots, nor prove the heresy of Copernicanism. He was never tortured by agents of the Roman Inquisition, nor languished in its dungeons. Nor did he ever mutter the phrase "E pur si muove" (And yet it moves), after being forced to recant his belief that the Earth moved around the Sun.

Galileo was no martyr of science.

Yet all these claims have been made, either by Galileo himself, or on his behalf. And when Newton declared, "If I have been able to see farther, it was because I stood on the shoulders of giants," one of those "giants" was undoubtedly Galileo.

So, was Galileo a Machiavelli of science? And what contribution did he actually make to the scientific revolution that was so great that, despite his legendary and infamous conflict with the Holy Roman Empire, he eventually became one of the Italian glories, to lay forever in the Basilica di Santa Croce, a Basilica of Roman Catholicism and the largest Franciscan church in the world?

THE SCIENTIFIC REVOLUTION

By the fifteenth century, the development of trade was gaining momentum. As the old economy of feudalism plodded along, a new order in economy and science finally began to surface. Improvements in technique, transport, and trade helped the emerging merchant class of burghers, or bourgeoisie, to transform the economy from one based not on service but on money.

The transformation had already begun in thirteenth century Italy. But it was not until the middle of the seventeenth century, even in the most progressive states of Holland and Britain, that the capitalist economy took a firm hold. Two further centuries were to pass before the merchants had control of the whole of Europe.

Through this entire period, 1450–1690, experimental science grew apace with capitalism. The triumph of both was far from easy. Ultimately, it was secured only after the most severe intellectual, religious, and political conflicts. At first, the conditions of the rise of capitalism gave rise to the new experimental method. By the end of this period, the reverse effect had come into play; experimental success in science led the next great technical advance, the Industrial Revolution.

The change was a complex one. Advances in technique led to new science. And science in turn led to new and accelerated changes in technique. Indeed, this combined scientific, technical, and economic revolution is a unique social process, a process that was ultimately greater in importance than even the discovery of agriculture.

The whole period of the scientific revolution was, of course, a single, unfolding transformation from a feudal to a capitalist economy. But to help clarify an understanding of the contribution and context of astronomers such as Copernicus and Galileo, it is useful to conceive of the revolution as having a backdrop of three distinct phases: the Renaissance, 1440–1540; the Wars of Religion, 1540–1650; and the Restoration, 1650–1690.

The parallel progress in science was to turn the world upside down.

In the first phase, the entire classical world picture was confronted. The highest expression of this conflict was that of Copernicus. He was to confront the geocentric cosmos of Aristotle and trade it in for a solar system seen from a revolving Earth, a planet like all the rest.

In the second phase, the change was made to tell. With the likes of Galileo and the new experimental method, science overthrew the worldview of the Christian schoolmen. The new world picture that emerged was quantitative rather than qualitative, atomic rather than continuous, infinite rather than bounded, and secular rather than religious.

By the end of the third phase, the success of science was complete. The world machine of Newton triumphed over the hierarchical universe of Aristotle.[2] This watershed signals the point at which final causes gave way to mechanical means, a universe of atomic particles interacting freely, guided by natural rather than divine law.

THE RENAISSANCE

A new economy dawned, and with it, the Renaissance.

Through Italy, High Germany and the Low Countries, an economic mode of commodity production and money payments developed. Only in Italy did this develop in the greater cities. Florence and Venice, Milan and Genoa, became politically and economically independent cities. The artistic and intellectual brilliance of the Renaissance began to bloom.

At first, The Holy See was happy. For Rome made a handsome profit from all of Christendom, Renaissance cities included.

In Germany there was war. The new economy inspired an independence of religion, the Lutheran Reformation, the social strife associated with the Peasant's War of 1525–1526, and the revolt of the Anabaptists between 1533–1535. In time the Reformation would spread further, bringing down the entire edifice of the hierarchical Church order.

Nation states began to emerge. Courts of kings or princes rested their power on the support of the merchants. Patronage was provided for new humanists and scientists, no longer dependent on the bishops. Outside of Italy, the old medieval universities remained the stronghold of feudal ideas, a bastion of Aristotelianism, and a stalwart opposition to the new learning.

The Renaissance and the Reformation became in many ways two expressions of the same struggle: the campaign to change the hierarchical system of fixed hereditary status and transform it to one based on the exchange of commodities and labor. Technical improvements had produced a trade surplus that led to rapid and vigorous expansion. And the availability of the surplus helped hugely with the speedy progress in navigation and shipping.

Navigation led to short circuits, and to booty. The old and costly land-based trade routes were replaced by swift and cheaper wet ones, the most spectacular being the New World of America. And the greed for more profit and booty fed back into the drive for shipbuilding and navigation. Technical advance was far more easily disseminated through the recent introduction of printing, as printed books appeared on subjects such as agriculture, cooking, and trade.[3]

And key to all this was the spirit of revolution in the air.

In science, art, and society, a conscious vanguard of merchants, scholars, and artists set about the task of constructing a new culture, capitalist in its economy, classical in its art, and scientific in its approach to nature.[4] At first, the plan had merely been to reconstruct the classical world, going back to the original Plato and Aristotle, the genuine Democritus and Archimedes. But soon the discovery of the New World made the classical world seem provincial.

The humanist Jean Fernel, physician to the King of France, expresses this new spirit in his *Dialogue* of 1530:

> But what if our elders, and those who preceded them, had followed simply the same path as did those before them? ... Nay, on the contrary it seems good for philosophers to move to fresh ways and systems; good for them to allow neither the voice of the detractor, nor the

weight of ancient culture, nor the fullness of authority, to deter those who would declare their own views. In that way each age produces its own crop of new authors and new arts.

This age of ours sees art and science gloriously re-risen, after twelve centuries of swoon. Art and science now equal their ancient splendour, or surpass it. This age need not, in any respect, despise itself, and sigh for the knowledge of the Ancients ... Our age today is doing things of which antiquity did not dream ... Ocean has been crossed by the prowess of our navigators, and new islands found. The far recesses of India lie revealed. The continent of the West, the so-called New World, unknown to our forefathers, has in great part become known.

In all this, and in what pertains to astronomy, Plato, Aristotle, and the old philosophers made progress, and Ptolemy added a great deal more. Yet, were one of them to return today, he would find geography changed past recognition. A new globe has been given us by the navigators of our time.[5]

The humanistic movement had begun in Italy as early as the fourteenth century. It was embodied by a dismissal of feudal ideas of hierarchy and a more secular approach to society. The cult of the individual became the ideal.[6] The old feudal culture had been found wanting, and a radical new system of thought had arisen, built up as part and parcel of a new social system by the actions of men who were making the revolution.

As is perfectly exemplified by the great engineer, scientist, and artist, Leonardo da Vinci, technicians and artists were no longer despised, as they were in the classical world. Art and architecture burgeoned. But so did the practical arts such as pottery, spinning and glass making. The might of the miners and metal workers was acknowledged, if only because of their key role in the drive to wealth and war.

The craftsman was married to the scholar. The improved status of such craftsman enabled the renewal of the link between the two for the first time since the early civilizations. And both would benefit from the relationship of knowledge and action. As Galileo announces at the very beginning of one of Brecht's greatest plays:

Times are changing, and we now have a new age. In Siena, when I was a young man, I saw some builders, after an argument lasting barely five minutes, discard a centuries old method of moving granite blocks in favour of a new way, a simple rearrangement of the ropes. It was then that I knew that the new age had arrived. What the old books say just isn't good enough anymore.[7]

THE COPERNICAN REVOLUTION

The fusion of knowledge and action was explosive.

The main intellectual mission of the Renaissance was to rediscover and master nature. And its greatest achievement was a system of the heavens with the Sun at the center. The picture that Nicholas Copernicus published in *De Revolutionibus Orbium Coelestium* (*On the Revolutions of the Celestial Orbs*, 1543) was clear enough for all to see. It was a system of the new lay society. Sure, it was based, at least in part, on the ancient learning. But there was a big difference. The new learning would be seen and experienced for itself. At first, the political repercussions of the new learning were not evident. Only with Galileo did the Church take fright and tried, too late, to shut it out.[8]

It is no wonder that it was astronomy, so closely related to navigation, that brought the whole ancient system of thought crashing down. The economic impact of the great voyages was decisive, not just on trade, but also on science. Economically, the navigations had inspired a trade comparable with the old internal trade of Europe. Scientifically, the greed for more profit led to a further rapid development of navigation, for an astronomy accurate enough to lead the agents of empire to safe and lucrative shores.

The lure of booty was key. The enormous demand for shipbuilding and navigation led to the establishment of schools of navigation in Spain, Portugal, England, France, and Holland.[9] A new class of craftsmen emerged. Intelligent and mathematically trained, they were adept in the skills of the compass, cartography, and instrument making.

The motion of the stars had cash value.

Indeed, the piracy that raged at this time on the high seas, itself indicative of the conflicts over trade and colonization among rival European powers, often sought after an unexpected booty. If a maritime raid proved successful, the boarding pirates would head straight for the hold. For rather than gold, silver, or pieces of eight, the most valuable cargo a ship possessed would be its maps and chronometer. Some cartographers would deliberately include errors on their maps, aiming to mislead the uninitiated should the map get into the wrong hands.

Understandably then, astronomy became the grindstone upon which the new science would gradually be sharpened. This first phase of the scientific revolution was qualitative, descriptive, and critical, rather than quantitative and constructive. That was to come later.

So it was with Copernicus. His clear and detailed explanation of a rotating Earth in a Sun-centered system is a decidedly descriptive astronomy, somewhat devoid of observation and experiment. Nonetheless, Copernicus was possessed of the new critical spirit.

The idea of a rotating Earth was nothing new, of course. As we have already seen in Chapter 3, in the third century BC Aristarchus had not only pictured an Earth in rotation, he had set out the first heliocentric system. His model had remained dormant ever since, a paradoxically alternative and absurd cosmology that suggested it was the Earth that moved, even though it was patently the Sun, Moon, and stars that could be seen to do so.

Armed with the inspiration of newly edited texts, a capable humanist could balance one Greek authority against another: Pythagoras against Plato, Democritus versus Aristotle, Ptolemy against Aristarchus. Equipped with a strong aesthetic sense, and plenty of courage, such a humanist may even dare to center the universe about the Sun, and worry later about shifting the Throne of God.

The man who dared do so was Copernicus. He had the courage to confront common sense, a sufficient skill in science to make the Sun star of his system, and, as a Renaissance humanist, enough incentive to bring the whole edifice of ancient thought plummeting down.

Nicholas Copernicus was born in Toruń, Poland, in 1473, a mere generation into the Renaissance. He was born of the patrician class, his mother from a rich merchant family, his father a wealthy copper trader who had become a respected citizen of that city. Indeed, the Polish rendering of their surname, Kopernik, means "one who works with copper."

Copernicus was a man of his time. He received Renaissance training in Italy: educated in astronomy at Bologna, medicine at Padua, and law at Ferrara. He spent the majority of his life as a canon in Frauenberg, and for the latter part was a burgher within the Bishopric of Warmia. Frauenberg was a cathedral town and situated in a disputed territory. With the kingdom of Poland on one side, and the Teutonic knights on the other, Copernicus was often working with the offices of war.

Astronomy was both an escape and a fascination. He seems to have committed his entire private life to building his own vision of the cosmos. For the last fifteen years of his life, he refined his heliocentric theory. Terrified of the social, political, and religious repercussions his theory would inflame, however, the Copernican model was not set out in its final form until its creator lay on his deathbed.

Copernicus's book *On the Revolutions of the Celestial Orbs*, finally published in 1543, suggests a system of spheres in some ways just as complex and bizarre as the theories it was meant to replace. But there was one key difference. They centered on the Sun, and not the Earth. He may have made few observations, and his intentions may have been mystical rather than scientific, but it was Copernicus's spirit of innovation that counts.

In this way, Copernicus is to astronomy what Columbus was to navigation. Not the greatest exponent of the subject, but a man who had the courage to confront his convictions and follow them through. It is this same character that is at the very heart of the Renaissance, and one that marks such a decisive break from the aimless Middle Ages.

For the first time in written history, the Sun and its planets were set out in the correct order. But the main reasons for Copernicus's revolutionary change were philosophic and aesthetic.[10] Writing about his heliocentric system, Copernicus here dwells on its inference of the near infinite distance to the stars:

> I think it easier to believe this than to confuse the issue by assuming a vast number of Spheres, which those who keep Earth at the centre must do. We thus rather follow Nature, who producing nothing vain or superfluous, often prefers to endow one cause with many effects.[11]

After detailing each planetary orb in turn, Copernicus then concludes:

> In the middle of all sits Sun enthroned. In this most beautiful temple could we place this luminary in any better position from which he can illuminate the whole at once? He is rightly called the Lamp, the Mind, the Ruler of the Universe; Hermes Trismegistus names him the visible God, Sophocles' Electra calls him the All-seeing. So the Sun sits upon a royal throne ruling his children the planets which circle round him. The Earth has the Moon at her service. As Aristotle says, in his de Animalibus, the Moon has the closest relationship with the Earth. Meanwhile the Earth conceives by the Sun, and becomes pregnant with annual rebirth.[12]

So 1543 AD marks one of the great turning points in human history.[13] Copernicus's great book, *De Revolutionibus Orbium Coelestium*,[14] placed the Sun at the center of our planetary system, heretically downgrading the position of the Earth to that of mere planet. Copernicus set in train a revolution. A new physics was born, and a new mantra: if the Earth is a planet, then the planets may be Earths; if the Earth is not central, then neither is humanity.[15]

COPERNICANISM CATCHES FIRE: THE CASE
OF GIORDANO BRUNO

At first the world held its breath.

The appearance of the new "solar system" took time to impact, and Copernicus's revolution was held in suspended animation. A handful of progressive astronomers recognized its elegance and simplicity for refining calculations. The German astronomer and mathematician Erasmus Reinhold, the most influential astronomical pedagogue of his generation, produced the Prussian Tables based on the Copernican model in 1551.

But few people believed they were really true. For one thing, the entire affair was an affront to common sense. For another, it seemed inconceivable that the Earth could revolve and rotate without whipping up a mighty wind, or warping the fall of shot. They were objections that Galileo would shortly remove.

Soon the seduction started. The very notion of an open universe, one in which the Earth was a mere atom, was bound to shatter the old image of Aristotle's closed cosmos of crystalline spheres, divinely made and maintained in motion. The new learning had led to new worlds on Earth. Might there not also be new worlds in the sky?[16]

This was the heresy for which Giordano Bruno was to be burnt at the stake.

Born near Naples just five years after the publication of Copernicus's great book, Bruno was soon distinguished for his outstanding ability. During his time in Naples he became known for his skill with the art of memory, traveling to Rome to demonstrate his mnemonic system before Pope Pius V. But Bruno was also possessed of a penetrating imagination, and a fiery temperament.

Though he dazzled merchants and scholars alike, trouble was never very far away. His acerbic tongue meant he was always on the move. He fell out with the monastic order to which he belonged, and his travels began in earnest.

The itinerary of Bruno's journey reads like a magical mystery tour of Renaissance culture. Tutored privately at an Augustinian monastery in Naples, Bruno entered the Dominican Order at the age of seventeen. His taste for free thinking and forbidden books soon caused him difficulties. When his annotated copy of the banned writings of Erasmus was found hidden in the convent privy, Bruno fled, shedding his monastic habit.

His years of wandering began.

Having visited Savona, Turin, Venice, Padua, and the Genoese port of Noli, Bruno ended up in Geneva. He was gifted a pair of breeches, a sword, hat, cape, and other accessories, hardly fitting clothing for a priest. Things went well until Bruno's published attack on the work of a local and distinguished professor forced him to leave Geneva.

His travels continued. A doctorate in theology and a lectureship in philosophy at Lyon, a meteoric rise to fame in Paris, where his incredible feats of memory were attributed by some to his magical powers. Indeed, his published work on mnemonics, *De Umbris Idearum* (*On The Shadows of Ideas*, 1582), was dedicated to King Henry III of France. Given that sixteenth century dedications were approved in advance, it is clear that Bruno had begun to move in powerful circles.

On the recommendation of Henry III, and as a guest of the French ambassador, Michel de Castelnau, Bruno went to England in 1583. There he met with one of the Elizabethan Age's most prominent figures, the English poet Philip Sidney, to whom Bruno dedicated two books. Bruno also became friendly with members of the Hermetic circle. The circle was focused around sometime Welsh physician to Queen Elizabeth I, John Dee, another mystic who dangerously straddled the worlds of science and magic, just as they were becoming distinguishable.

Bruno's dabbling in Copernican cosmology began in earnest. While lecturing at Oxford, he failed in his attempt to secure tenure there, but was successful in courting controversy with potentially powerful priests. Bruno's spat involved John Underhill, Rector of Lincoln College (and, from 1589, Bishop of Oxford), and George Abbot who was soon to become Archbishop of Canterbury. It was Abbot who derided Bruno for holding "the opinion of Copernicus that the earth did go round, and the heavens did stand still; whereas in truth it was his own head which rather did run round, and his brains did not stand still."[17]

Nonetheless, Bruno's stay in England was a productive one, publishing some of his most important works. Among these his *Italian Dialogues* is significant, for they include cosmological tracts such as *La Cena de le Ceneri* (*The Ash Wednesday Supper*, 1584), *De la Causa, Principio et Uno* (*On Cause, Principle and Unity*, 1584), and *De l'Infinito Universo et Mondi* (*On the Infinite Universe and Worlds*, 1584).

Bruno believed in aliens.

His *On the Infinite Universe and Worlds* was a form of cosmic pantheism, a belief that the universe and God is one and the same thing. Bruno was passionately pluralist. He populated planets and stars, attributing them souls, and even did the same for the universe as a whole. In Bruno's words, "Innumerable suns exist; innumerable earths revolve around these suns in a manner similar to the way the seven planets revolve around our sun. Living beings inhabit these worlds."[18]

As another great champion of free thought, Robert Green Ingersoll, the American political leader, was to say of Bruno:

> The First Great Star, Herald of the Dawn, was Bruno ... He was a pantheist, that is to say, an atheist. He was a lover of Nature, a reaction from the asceticism of the church. He was tired of the gloom of the monastery. He loved the fields, the woods, the streams. He said to his brother-priests: Come out of your cells, out of your dungeons: come into the air and light. Throw away your beads and your crosses. Gather flowers; mingle with your fellow-men; have wives and children; scatter the seeds of joy; throw away the thorns and nettles of your creeds; enjoy the perpetual miracle of life.[19]

With Bruno, the true spirit of Copernicanism had also taken flight. His favored cosmology was far more radical than that of Copernicus. A rotating and revolving Earth was responsible for the illusion of the night sky. The stellar realm was infinite in extent, and Bruno saw no reason to believe that the stars were equidistant from a single center of the universe.

Having published over 120 theses against Aristotelian natural science, Bruno dramatically and completely abandoned the idea of a hierarchical universe. The Earth was just one more heavenly body, as was the Sun. God had no more business with the Earth than he did with any other far-flung planet. According to Bruno, God was as present in the heavens as he was on Earth, an Immanence, subsuming the great multiplicity of existence, instead of the rather pitifully remote heavenly deity preferred by the Church.

The universe of Bruno was the same universe of the Atomists, an infinite universe, in a constant state of flux. Space and time had no end, and the same physical laws governed every last corner of the cosmos, which was made up everywhere of the four elements, rather than having the stars composed of a different quintessence. In this vision, there is no room for stability and permanence, for Christian notions of divine creation and a Last Judgement.

In Bruno's scheme of things, the Sun was simply a star, not the high-flown mystic body of Copernicus's vision. All the stars were suns, and the cosmos was composed of an infinite number of solar systems, the basic building block of the entire universe. Matter was made of discrete atoms, separated by vast realms full of ether, since Bruno believed empty space could not exist.

His ideas were held up for ridicule. After going back to Padua, he taught briefly, and then applied unsuccessfully for the chair of mathematics, a post that was to be assigned instead to Galileo, one year later. After his last years of wandering, The Inquisition finally summoned Bruno to Rome. There, he was incarcerated for seven years during his drawn-out trial, lastly, in the Tower of Nona and the ill-famed dungeon's terrible lightless cells, one of which was known as "the pit." Some key papers about the infamous trial are lost, but a stunning find was made as recently as 1940, when the summary of the trial's proceedings was surprisingly unearthed.[20]

Bruno was stitched up by the Inquisition. The charges included blasphemy, heresy (of course), and immoral conduct. But, most importantly for our story, especially in view of what was to come with Galileo, Bruno was also charged on the basis of his doctrines of philosophy and cosmology.

The charges included claiming the existence of a plurality of worlds and their eternity, holding opinions contrary to the Catholic Faith and speaking against it and its ministers, holding erroneous opinions about the Trinity, holding erroneous opinions about Christ, denying the Virginity of Mary, and dealing in magics and divination.

Faced with this bleak state of affairs, Bruno courageously set out his defense. Crucially, he was prepared to bow to the Church on its dogmatic teachings, but held firm on his philosophy. Above all, his Copernican belief in the plurality of worlds he held most dear, even though he was told to abandon it.

At length, on February 10, 1600, Bruno was led out to the Church of Santa Maria, and sentenced to be burnt alive. As the Holy Church was to curiously phrase it, he was to be punished "as mercifully as possible, and without effusion of blood." In insolent response, Bruno famously replied, "You are more afraid to pronounce my sentence than I to receive it."[21]

He was allowed a week's grace for recantation, to no avail. On February 17, 1600, Bruno was burnt to death on the Field of Flowers. To the end he was brave and rebellious. The story goes he contemptuously pushed aside the crucifix they gave him to kiss. As one of his enemies admitted, he died calm and heroic, defiant of wrong.

The secularist and freethinker George William Foote found great inspiration in Bruno's example.

Such heroism stirs the blood more than the sound of a trumpet. Bruno stood at the stake in solitary and awful grandeur. There was not a friendly face in the vast crowd around him. It was one man against the world. Surely the knight of Liberty, the champion of Freethought, who lived such a life and died such a death, without hope of reward on earth or in heaven, sustained only by his indomitable manhood, is worthy to be accounted the supreme martyr of all time. He towers above the less disinterested martyrs of Faith like a colossus; the proudest of them might walk under him without bending.[22]

Controversy still rages over both the reasons for Bruno's death and his contribution to the scientific revolution.

His ideas were at first held up for ridicule. English aristocrat and prolific writer Margaret Cavendish, for example, penned a run of poems against "atoms" and "infinite worlds" in her *Poems and Fancies* in 1664. But Bruno's potential began to shine through with the implications of Newtonian cosmology, an infinite universe lacking in structure, and a crucial cross point between the old and the new.

Arguably, and far more convincingly than Copernicus, it is the cosmology of Bruno that first marks the paradigm shift of the old universe into the new. Aristotle's cozy geocentric cosmos was about us. The new universe of Bruno was decentralized, inhuman, and alien. Now we can more clearly see the cultural shock created by the discovery of humanity's marginal position in a universe fundamentally inhospitable to man.[23]

Certainly, Bruno's universe is strikingly modern. It is infinite, homogeneous, and isotropic, or uniform in all directions. Planetary systems are now distributed evenly throughout the cosmos. It is even possible to see in Bruno's idea of multiple worlds, and the infinite possibilities of the indivisible One, a forerunner of the many-worlds interpretation of quantum mechanics.

As to Bruno's death, there are clear parallels with the case of Galileo.

For one thing, many see Bruno as the true and original martyr of science, identifying his Copernicanism as a significant factor of the outcome of his trial. As John J Kessler writes, "He is one martyr whose name should lead all the rest. He was not a mere religious sectarian who was caught up in the psychology of some mob hysteria. He was a sensitive, imaginative poet, fired with the enthusiasm of a larger vision of a larger universe ... and he fell into the error of heretical belief."[24]

For another, the inquisitor Cardinal Bellarmine, the "Hammer of the Heretics," oversaw Bruno's trial. It was Bellarmine who demanded a full recantation, which Bruno eventually refused, and, as we shall soon see, Bellarmine was to play a crucial role in the Galileo affair.

There are some who hold that to connect Bruno's death with his cosmology is entirely mistaken. To support this view, they point out that in 1600 the Catholic Church had no official position on the Copernican system, and that it was certainly not heresy.[25] Copernicanism, they suggest, was not specifically proscribed as heretical until well after Bruno's death, when Bellarmine himself placed Copernicus's *De Revolutionibus*, among others, on the "Index of Forbidden Books."

But there is far more in heaven and hell than the documentary evidence of a willful bureaucracy. Besides, the secret archives of the Vatican show a different story. In the detail of Bruno's trial in Rome, the documents state:

> In the same rooms where Giordano Bruno was questioned, for the same important reasons of the relationship between science and faith, at the dawning of the new astronomy and at the decline of Aristotle's philosophy, sixteen years later, Cardinal Bellarmino, who then contested Bruno's heretical theses, summoned Galileo Galilei, who also faced a famous inquisitorial trial, which, luckily for him, ended with a simple abjuration.[26]

For his belief in aliens, along with other alleged heresies, the Roman Catholic Church burned one Italian philosopher in Giordano Bruno at the stake in 1600. He certainly suffered a cruel and unnecessary death. Some say it was a martyr's death. Others say that Bruno, and not Galileo, has become the Church's most difficult alibi.

> The murder of this man will never be completely and perfectly avenged until from Rome shall be swept every vestige of priest and pope, until over the shapeless ruin of St. Peter's, the crumbled Vatican and the fallen cross, shall rise a monument to Bruno, the thinker, philosopher, philanthropist, atheist, martyr.[27]

It is time to consider the Galileo affair.

THE TELESCOPE

Perhaps one of the few failings of Bruno's cosmology was his belief that the universe was essentially animistic, and that the motion of matter was due to the movement of the souls and spirits of which matter was made. His approach also seems to have possessed a corresponding disregard for mathematics as a means of understanding. It is the most dramatic way in which Bruno's cosmology differs from today's state-of-the-universe picture.

The invention that was to prove crucial for Copernicanism, and the acceptance of the new worldview, was the spyglass. Rather than any further critical refinement to theory, it was the inception of the telescope, the *far-seer*, that was to bring the heavens down to Earth, and Aristotle's cosmology to its knees.

The technology that made up the telescope had long been familiar. But the history of optics is another fascinating episode of the social relationship between theory and technique. Ancient civilizations had noticed the curious lens effect produced by, say, transparent crystals, glass spheres full of water, or jewels. Indeed, the magnifying effects were something of a fascination.

But it is as though those observations came too soon. In the theories and texts of many ancient Greek and particularly medieval Arab scholars, it is clear that a theory of optics was in existence. The technology to magnify had also been available. And clearly the demand for the technology of magnification in commerce and warfare was legion. But theory was never married to practice, never tallied up to the optical devices that may have been available.

This curious state of affairs can be explained in two related ways. First, a social explanation. It is very unusual in history, as we have seen, for scholars and craftsmen to work together, given their difference in social class. This was the case until the European Renaissance, one of the distinguishing features of which was the intensity and distribution of the collaboration between the work of the mind and the work of the hand.

Second, the old adage "a little knowledge is a dangerous thing" may also apply here. Ancient or medieval observations made through transparent crystals, glass spheres, or jewels, do not just show things bigger, or better. They can also deceive. To the more superstitious mind, a mind not married to rational theory, such optical illusions can make one wary. Sight is the most paradoxical of senses. It is at once the most reliable, and yet the most deluding. Without theory, perhaps you can not really trust what you see.

Ultimately, the Islamic scholars Ibn Sahl and Ibn al-Haytham had, in the tenth and eleventh centuries, refined the technical knowledge necessary for the production of spectacle lenses. The Islamic findings soon found fashionable favor in Europe. Florentine Salvino D'Armate is credited with inventing the first wearable eyeglasses around 1284. But the evidence is contentious. The claims to D'Armate's primacy are merely based on his now defunct memorial inscription, "Here lies Salvino degl' Armati, son of Armato of Florence, inventor of eyeglasses. May God forgive his sins." However, the pictorial evidence is incontrovertible, including paintings such as Tomaso da Modena's 1352 portrait of the cardinal Hugh de Provence, reading in a scriptorium.

Written evidence survives that the principle of the telescope was known to English surveyor Leonard Digges,[28] Muslim polymath Taqi al-Din,[29] and Neapolitan scholar Giambattista della Porta,[30] all in the late sixteenth century. Alas, no designs or physical evidence survives.

Unsurprisingly, Leonardo da Vinci also reflected on the possibilities of the telescope in his notebooks between 1508 and 1510: "It is possible to find means by which the eye shall not see remote objects as much diminished as in natural perspective," was his view in one notebook;[31] and more specifically in another: "The further you place eyeglasses from the eye, the larger the objects appear in them … And if the eye sees two equal objects in comparison, one outside of the glass and the other within the field, the one in the glass will seem large and the other small."[32]

But, the earliest known working telescopes were lens-based refracting telescopes that materialized in the Netherlands in 1608. Their development is attributed to three separate Dutch spectacle makers: Jacob Metius of Alkmaar and Zacharias Janssen and Hans Lippershey from Middelburg. Legend has it that a couple of children playing with lenses in Lippershey's shop first gazed through one lens at another in the shop window, around the year 1600. Curiously, the blend of lenses made things outside the shop seem closer.

The very fact that no scientist was needed to invent the telescope is worthy of note. The device was long overdue, and the means of making a spyglass had probably been present for at least three hundred years. Only with the sheer volume of lens manufacture through the increased wealth of the sixteenth century was the discovery brought about by chance.

GALILEO GALILEI

The telescope became the Renaissance's weapon of discovery. Its invention and use in anger for science was to signal the true offensive on the feudal order of the old universe. By the end of the century, the layered universe of Aristotle had crumbled before the world machine of Newton. And as Newton himself later admitted, his vision was made all the more cloudless by the gigantic contribution of Galileo.

Galileo did not invent the telescope. It is Hans Lippershey who is credited with both creating and distributing designs for the first "optick tube." Lippershey sent his patent to the Estates General of the Netherlands for a license to design and make single and double lens scopes for thirty years, making it available for general manufacture and use.

Lippershey never did receive a patent, because in the meantime Metius and Janssen had also claimed the invention of the spyglass. But his government generously rewarded him for the design of the "Dutch perspective glass," gifting two of Lippershey's optic tubes to the King of France. By April 1609, Parisians could splash out on their own spyglass, from spectacle-makers' shops in the city.

In the same year, the telescope found its way to Italy. Galileo was later to claim that he built his own gadget, based on the reports he read of the Dutch invention, and after he had studied the theory of refraction. This is perfectly possible. For lesser minds than Galileo's had certainly already done so. But Galileo managed to refine the design. The "Dutch perspective glass" was only capable of three times magnification. The magnification of Galileo's first prototype was a full three times more powerful.

Galileo wasted no time in realizing the potential of the spyglass for trade and war. He had occupied the position of chair of Mathematics at the famous University of Padua since 1592, the year after Bruno had unsuccessfully applied for the same post. Now, on August 8, 1609, Galileo arranged a meeting with the powerful Venetian Senate.

The meeting with the Senate took place at St. Mark's Campanile, the bell tower of St. Mark's Basilica, the most familiar symbol of Venice. The tower affords panoramic views across the Venetian skyline, city, and sea. Members of the Senate examined Galileo's spyglass with great curiosity and excitement. The nine times magnification made it possible to see "sails and shipping that were so far off that it was two hours before they were seen with the naked eye, steering full sail into the harbour."[33]

The meeting was a resounding success. By August 11, 1609, Galileo had made a gift of his telescope to the Senate, taking care to point out that since he had illustrated how invaluable the gadget was against invasion by sea, it would prove to be of supreme strategic importance in war.

It was a shrewd and calculated move. The Venetian Senate, indebted to Galileo for the gift of the gadget, immediately doubled his annual salary to a thousand scudi. His professorship at Padua, a university under the sway of the Venetian Republic, was granted lifelong tenure. Within weeks, local spectacle makers had managed to make telescopes of the same magnifying power. On the streets of Venice, they were selling for a few scudi, the gadget that Galileo had flogged to the Senate for a thousand a year. The locals were greatly amused.

But the move becomes even more shrewd and calculated. Just days before Galileo's celebrated meeting with the Senate, Hans Lippershey had arrived in Venice to show his own device to the city fathers. Along with his friend Paolo Sarpi, Galileo had contrived to block Lippershey. As Sarpi bought Galileo time by obstructing the potential business meetings of their rival, Galileo labored away, refining his own telescopic design.

Our first encounter with Galileo presents the immediate impression of a man with a keen sense of publicity, of the great material value of his work, which he did not seem to think incompatible with the élan of his discoveries. In this sense, Galileo seems thoroughly modern.

He had been born in Pisa in 1564, the same year as Shakespeare, dying in 1642, the year Newton was to be born. Galileo's father, Vincento Galilei, was a composer of considerable accomplishment. A member of the lower nobility, his radical leanings and contempt for authority seems to have been easily handed down to his eldest child. Like Bruno before him, Galileo seems to have been quick of temper, acerbic of tongue, and intolerant of opposition.

Since his early twenties, Galileo had been a convinced Copernican. So it is no surprise that now, at the age of forty-five, and with the prospect of delving into the deep, Galileo turns his spyglass at the heavens. First, he began an intense period of labor on his own design of optic tube. Then, armed with his greatly refined device, over the next few months he proceeded to make observations that systematically laid bare the outrageous nonsense of Aristotle's cosmology.

THE STARRY MESSENGER

In March 1610, Galileo's revolutionary discoveries with the spyglass were hurled like an incendiary device into the arena of a dull academic world. His evidence was laid out in the shape of a brief but magnificent flyer, the momentous and now legendary book, *Sidereus Nuncius (The Starry Messenger)*.

The book itself, like the content of its brief twenty-four pages, was radical. Incredibly, *The Starry Messenger* was Galileo's first scientific publication. But not only did it tell of discoveries "which no mortal had seen before,"[34] it was also penned in a new, tersely written way, which no scholar had previously used. The first print run was of 550 copies, sent out to key figures in the universities of Italy, as well as to booksellers in every major city.

It was a revolution in the communication of science. The result was culture shock. The sophisticated Imperial Ambassador in Venice said the book was "a dry discourse or an inflated boast, devoid of all philosophy."[35] Nonetheless, the book created the dramatic effect for which it was designed, arousing immediate and passionate controversy.

The Starry Messenger begins with Galileo's account of how he refined his telescope, "by sparing neither labour nor expense, in constructing for myself an instrument so superior that objects seen through it appear magnified nearly a thousand times, and more than thirty times nearer than if viewed by the natural power of sight alone."[36]

In great contrast to the ancients, who felt degraded by servile labor, Galileo had no qualms about combining the work of the mind with the work of the hand. Such was the revolutionary effect of the Renaissance, where scholars and craftsmen worked together. Armed with his spyglass, and after the introduction in *The Starry Messenger*, Galileo goes on to conclude his observations of the Moon:

> The surface of the Moon is not perfectly smooth, free from inequalities and exactly spherical, as a large school of philosophers considers with regard to the Moon and the other heavenly bodies, but that, on the contrary, it is full of irregularities, uneven, full of hollows and protuberances, just like the surface of the Earth itself, which is varied everywhere by lofty mountains and deep valleys.[37]

The fascinating passage above makes Galileo's intentions clear. Aristotle had been the last notable ancient cosmologist, his vision casting a long and dark shadow over astronomy for eighteen centuries. But Galileo shows how Aristotle's system does not stand up to modern scrutiny.

Consider the evidence, he says. In Aristotle's two-tier cosmos, only the Earth was subject to the horrors of change, death, and decay. Beyond the Earth, the celestial sphere was immutable and perfect. And yet through the telescope, the picture is different. The Moon is pock-marked and irregular, far from crystalline and perfect. As to Aristotle's claims that the Earth is unique, Galileo shows that the Moon has Earth-like mountains, valleys, and hollows.

Not only that, but anyone with enough ready cash could pop out to the local spectacle maker, pick up a spyglass, and see the surface of the Moon with their own eyes. Galileo himself had spent almost every night of the first three months after he built the device assiduously studying the lunar surface through his scope.

Next, in *The Starry Messenger*, Galileo turns his gadget to gaze at the stars, describing how the spyglass prises open the heavens for deeper exploration. We can imagine the night sky in the seventeenth century. Before these days of light pollution, the dark canopy of the night, bejewelled with stars, must have been a stirring sight to the naked eye. But even by those standards, and even though the optic-tube was far from perfect, the transformation was stunning.

The telescope reveals many "other stars, in myriads, which have never been seen before, and which surpass the old, previously known stars in number more than ten times."[38] To the nine stars in the belt and sword of Orion, Galileo added eighty more; to the seven stars in the constellation of Pleiades, the seven sisters, he added a further thirty-six; and as for the belt of the Milky Way, the telescope resolved the Galaxy into "a mass of innumerable stars planted together in clusters."[39]

A question in the nerves is immediately lit, of course. If the magnificent splendour of this God-given universe of Aristotle's is made especially for Man's delight, why is it only through a machine, such as the spyglass, that Man can justly savor its intricate parts, and begin to know its true nature?

Finally, Galileo unveiled the greatest revelation of all:

> There remains the matter which seems to me to deserve to be considered the most important in this work, namely, that I should disclose and publish to the world the occasion of discovering and observing four Planets, never seen from the very beginning of the world up to our own times, their positions, and the observations made during the last two months about their movements and their changes of magnitude; and I summon all astronomers to apply themselves to examine and determine their periodic times, which it has not been permitted me to achieve up to this day ... On the 7th day of January in the present year, 1610, in the first hour of the following night, when I was viewing the constellations of the heavens through a telescope, the planet Jupiter presented itself to my view, and as I had prepared for myself a very excellent instrument, I noticed a circumstance which I had never been able to notice before, namely that three little stars, small but very bright, were near the planet; and although I believed them to belong to a number of the fixed stars, yet they made me somewhat wonder, because they seemed to be arranged exactly in a straight line, parallel to the ecliptic, and to be brighter than the rest of the stars, equal to them in magnitude ... When on January 8th, led by some fatality, I turned again to look at the same part of the heavens, I found a very different state of things, for there were three little stars all west of Jupiter, and nearer together than on the previous night.[40]

He had at first noticed three objects he believed to be stars, close to Jupiter. During further nightly vigils, these three were joined by a fourth, and from the observed changes in both position and brightness, Galileo became convinced that the bodies were in orbit about

Jupiter: "Above all, since they sometimes follow and sometimes precede Jupiter by the same intervals, and they remain within very limited distances either to the east or west of Jupiter, no one can doubt that they complete their revolutions about Jupiter and at the same time effect altogether a twelve-year period about the centre of the universe."[41] In short, the soon to be known Galilean moons orbit Jupiter and the whole system orbits the center of the cosmos, be it Sun or Earth, every twelve years.

The impact and importance of this last sighting was not lost on the rather garrulous Galileo. The four new planets were the first four moons of Jupiter, their significance to the case of Copernicanism Galileo now sets out: "Moreover, we have an excellent and exceedingly clear argument to put at rest the scruples of those who can tolerate the revolution of the planets about the Sun in the Copernican system, but are so disturbed by the revolution of the single Moon around the Earth while both of them describe an annual orbit round the Sun, that they consider this theory of the universe to be impossible."[42]

It is the only name check for Copernicus in the entire twenty-four pages. But it is enough. Galileo's point is pitted against the critics of Copernicus who see problems with the Moon's composite motion about the Earth. Galileo makes such a motion seem commonplace, since his telescope evidences four moons keeping composite company with Jupiter.

Curiously, *The Starry Messenger* does not contain obvious and vulgar arguments in the cause of Copernicus. Rather, heliocentrism is a subtle subtext within its pages, coded in like a cipher, only to be evinced from the telescopic evidence that Galileo describes. This should not be surprising. Galileo's first ringing public assertion in favor of the Copernican system was not to come for another three years. During his time as professor at Padua, he taught the ancient system of Ptolemy and Aristotle. He even clearly contradicted Copernicus.

What had made Galileo previously so wary of publicly stating his true opinions on astronomy? There are at least three answers to this question. On the one hand, Galileo was no doubt aware of the fact that he was defending an incomplete thesis, with little or no evidence, at least until the telescope came along. On the other hand, like Copernicus, Galileo feared the potential ridicule from his fellow professors—to the last man defenders of the systems of Aristotle and Ptolemy.

And then there is the question of Bruno. It was probably the example of his terrible fate that convinced Galileo more than anything else to tread shrewdly and softly on the pernicious error of heliocentrism.

Indeed, there is some speculation as to whether Galileo and Bruno ever met. At Padua, Galileo enjoyed a good rapport with city noble-man and intellectual Gianvincenzio Pinelli. It is perfectly possible that Bruno and Galileo may have been on the guest list for dinner at Pinelli's luxurious home on the Via del Santo. Pinelli had been a consummate host, demanding that Galileo reside at the palazzo when Galileo had accepted the chair at Padua in 1592.

Another notable inclusion on Pinelli's guest list at that time would have been Robert Bellarmine, the Hammer of the Heretics mentioned in dispatches when we spoke of the case of Bruno. One can only imagine the topic of conversation between Galileo, Bruno, and Bellarmine over dinner. Bruno was famous across Europe as an antiestablishment figure. He played a dangerous game, and espoused what Bellarmine regarded as heretical views.

Like Galileo, Bellarmine was shrewd and very clever. He had an intense hatred for Bruno. But he waited, and watched. Bruno was not to die at the stake, as a corrupter of the Faith, for another eight years, a particularly protracted and terrible "witch-hunt," that must surely have left its mark in the calculated mind of Galileo.

THE COPERNICAN PROTAGONIST

Galileo's use of the spyglass in the name of science was not the first. English astronomer and mathematician Thomas Harriot is also known to have made observations with a telescope, in the summer of 1609. Indeed, he is even said to have made maps of the Moon, aided by a six-powered spyglass he developed about the time that Lippershey began to promote his own invention. And yet, Harriot did not publish, nor did he realize the mercantile possibilities of the device.

Galileo did both. Galileo's star maps may have been inaccurate, and some of the lunar features he describes do not actually exist, but the impact of *The Starry Messenger* was enormous. Even though his explicit references to Copernicus and his system are limited in the text, Galileo's first publication marks the watershed of his becoming a protagonist for Copernicanism.

The cumulative impact of the sightings is key.

The Moon is shown to be like the Earth, the presence of moun-tains and valleys on both suggesting the similarity of heavenly and earthly matter, a homogeneity of cosmic material. The myriad of stars unseen to the naked eye made mockery of the notion that God had created the universe fit for man's pleasure, when they could only be treasured through such trickery. And the moons in orbit around Jupiter exposed the ancient fallacy that the Earth was the sole center around which all revolved.

Only years before, the Danish astronomer and nobleman Tycho Brahe had begun to pick away at the ancient system. The world's most notable astronomer just before the days of the telescope, Tycho is credited with the most accurate astronomical observations of his time. Chief among these were two cosmic events that would begin to mark the death knell of Aristotle's cosmology.

On the evening of November 11, 1572, a new star had appeared in the constellation of Cassiopeia, the star system identified by its familiar "W" pattern in the night sky. The sight was so inconceivable that Tycho literally did not believe his own eyes. He summoned both servants and serfs to confirm that he was still of sound mind and that the nova was real.

The marvel was this: new stars simply were not allowed, at least not in the systems of Aristotle and the Church. Any change in the heavens, from the Moon to the sphere of the Prime Mover, had been effectively banned. All of Europe had been agog at the implications of this new star, and Tycho had published a book, *De Nova Stella*, in which he had nailed the lie of heavenly permanence, and with it the stability of the old, walled-in universe.

Five years later, Tycho delivered a further bombshell. The great comet of 1577 came soaring through western skies, veering close to Venus, and fetching the Sun as a compass. By showing the comet was "at least six times" further away than the Moon, Tycho proved Aristotle wrong. Comets were not atmospheric combustions after all. And if this particular comet really was soaring through interplanetary space, why could even the most casual observer not hear the crashing of the crystalline spheres as the comet made its way through the cosmos?

Now, armed with the telescope, Galileo had delivered the last rites on Aristotle's system. Copernicus's book *De Revolutionibus* had raised hardly a twitter. But with *The Starry Messenger*, Galileo became the champion of Copernicanism.

One key factor was the communication of science.

Copernicus had been so terrified of the religious response to *De Revolutionibus* that he had delayed publication until he lay on his deathbed. Even then the tome's preface had been stamped with the esoteric motto "For Mathematicians Only." Hardly a call to arms, and little surprising that the work came to be known as "the book that nobody read."

In great contrast, consider Galileo. He deliberately distributed *The Starry Messenger* far and wide. It was a very readable twenty-four pages long, easily read in an hour. And its resonant effect on the reader left them in little doubt that Galileo was calling time on the bounded universe of old.

The shock waves of Galileo's discoveries spread throughout the civilized world. It was an instant success in Italy, and copies were eagerly snatched from booksellers in Spain, England, and the Netherlands.

As is reported elsewhere,[43] on March 15, 1610, the rather improbably named Wackher von Wackenfels, privy counselor to his Holy Imperial Majesty, told the internationally famous scholar Johannes Kepler, of news that had just arrived at court: a philosopher named Galileo in Padua had turned a Dutch spyglass at the heavens, and discovered four new worlds.

Kepler wrote to Galileo enthusiastically and with great faith in his findings: "I may perhaps seem rash in accepting your claims so readily with no support of my own experience … but why should I not believe a most learned mathematician, whose very style attests the soundness of his judgement."[44]

Later in the same year of 1610, English poet John Donne turned his poetic talents to the sky. Donne was noted for a brilliant knowledge of English society, along with sharp satiric criticism of its problems. Now he turned his sights on Galileo, "Man has weav'd out a net, and this net throwne; Upon the Heavens, and now they are his owne … I will write [quoth Lucifer] to the Bishop of Rome: He shall call Galileo the Florentine to him."[45]

The news had hit England early. On the very day that *The Starry Messenger* was published, Sir Henry Wotton, author and British ambassador to Venice, had written to his home office describing the "strangest piece of news that hath ever yet received from any part of the world."[46] Wotton gives his own account of the importance of Galileo's discoveries with the telescope, including the "four new planets,"[47] the moons of Jupiter, and the news that the Moon is not spherical "but endued with many prominences"[48] and "illuminated with the solar light by reflection from the body of the Earth."[49]

Wotton then goes on to make a stunning prediction for his time. He considered that Galileo's sightings would turn the world upside down and revolutionize astronomy. For all of Europe was abuzz. "Here in Italy, all corners are full of talk about these discoveries,"[50] he gushed. His fascinated communication ends with the opinion that Galileo "runneth a fortune to be either exceedingly famous or exceedingly ridiculous."[51]

By the end of 1610, the author of *The Starry Messenger* was the hottest intellectual property in all of Europe. The English poet John Milton was but an infant when the discoveries were made. Later, in 1638, he would visit the Tuscan astronomer. With his eyesight still intact, Milton was not to know that, like Galileo, he would go blind, be incarcerated for his beliefs, and create his own revolutionary cosmos in his epic, *Paradise Lost.*

The Cromwellian Milton was to mature in full and sober awareness of the "vast unbounded Deep" unveiled by Galileo's telescope, reflecting the end of the medieval walled-in universe, in politics as well as in science:

> Before [his] eyes in sudden view appear
> The secrets of the hoary Deep—a dark
> Illimitable ocean, without bound,
> Without dimension.[52]

The revolution had only just begun.

CHAPTER 7

EVOLUTION AND DARWIN

If Galileo's discoveries with the telescope were a journey in space, Darwin's voyage on HMS *Beagle* was a trip through time. Darwin, like many scientists of the nineteenth century, was about to experience the ride of his life, a ride on which time itself would seem to warp and telescope. The only way in which he was unwittingly prepared for such a journey was in the form of a book he had packed for the *Beagle*'s voyage around the world.

The book was *Principles of Geology*, by the Scottish geologist Charles Lyell. Darwin's friend and botany professor at Cambridge, John Stevens Henslow, had recommended Darwin as the naturalist on the *Beagle*'s scientific survey of South American waters. He had also recommended Lyell's book, along with a quiet word of caution. Darwin should by all means enjoy Lyell's skills as a writer, but be wary of his radical conclusions. For Lyell was a uniformitarian. He believed that all change, in geology and biology, was down to slow, but unstoppable processes, running throughout Earth's long history.

The uniformitarian credo was a steady-state view of natural history. The slow attrition of soil and rock by wind and water, along with gradual changes in climate, were the processes responsible for the extinction of species to be found in the fossil record. James Hutton, the Scottish polymath, had been the first to propose Uniformitarianism: "The ruins of an older world are visible in the present structure of our planet."[1]

A schema, commissioned by Hutton, charmingly illustrates this uniformitarian thesis of sleepy subterranean rumblings. The cutaway drawing portrayed a pastoral idyll above ground, a covered carriage drawn through woods by two horses. But down below was a frozen image of a world in constant change: a frieze of strata, and underneath, a warped and twisted tableau of metamorphic rock. Forces that had the power to move mountains, in time.

The radical aspect of Lyell's *Geology*, and the reason that Darwin had been warned off coming to such similar conclusions, was in the question of time. Since uniformitarian processes acted slowly, on the whole, the theory required the Earth to be very old. Hutton estimated the extent of such antiquity rather succinctly: "We find no vestige of a beginning, no prospect of an end."[2]

No Genesis, and No Revelation

It was the prospect of such horror that led many fundamentalists to Catastrophism. The catastrophist response to the evidence of the fossil record was almost the opposite of the uniformitarian. Where the uniformitarians saw slow change as the way to explain natural phenomena, catastrophists saw sudden, near supernatural upheavals, levelling mountains in a minute, heaving up seabeds at a sweep, dooming entire species in a matter of seconds.

Crucially, Catastrophism explained extinction without flying in the face of Scripture. It entertained solid support from the biblical story of the Flood, which the catastrophists saw as one of many calamities visited to the Earth by a vengeful God. The catastrophists were divided on whether God still dabbled in disaster, however. George Cuvier, like many naturalists of the early nineteenth century, believed that, while the Flood may have flowed from a divine hand, modern catastrophes were down to more conventional actions.

Now, one of the main problems with Catastrophism was this: like the Greek idea of the "great year" that came before it, it is a theory that merely serves to sever the past from the present. In some ways the job of the geologist, in the early nineteenth century, was like that of Galileo, two centuries earlier. Galileo was faced with a schism: Aristotle's rifted cosmology of a mundane Earth and an immutable sky. But he had used the telescope to show that sunspots made the Sun changeable and impure, and that the Moon had mountains, just as the Earth. Galileo's discoveries implied one physics, heaven and Earth alike.

The new geology faced a similar schism. Catastrophism consigned geological change to the dustbin of history. The supernatural forces responsible for such change acted only in the early history of the Earth. Any telescoping of theory from the present into the past was essentially hopeless. As Lyell himself put it, "Never was there a dogma more calculated to foster indolence, and to blunt the keen edge of curiosity, than this assumption of the discordance between the former and the existing causes of change."[3]

The Christian interpretation of Earth history also faced problems on the biological front. Just as Galileo's telescope had unveiled a universe previously unknown and unseen, so the fossil record overflowed with a cornucopia of cadaverous creatures. Fossilized flowers that no one had ever seen in bloom, woolly mammoths and rhinos, the terrible lizards, and a whole host of bizarre beasts that looked like they had just walked out of Hieronymous Bosch's *Garden of Earthly Delights.*

And the remains of these extinct animals were found in curious locations, places where they could not possibly have prospered: sea creatures on hilltops, polar bears on the equator, and so on. The planet must have undergone massive upheavals, to say the least, for such profound changes to have happened in the short time imagined by the biblical literalists.

And if explaining away the visceral harvest of the fossil record was challenge enough to Christian theorists, there was also a mystifying array of living creatures to contend with. Through exploration, piracy, and plunder in the deep jungles of the globe, naturalists were faced with a burgeoning biota. Many beasts were so odious and obscure that it was hard to see why a divine hand would have introduced them into the world at all.

Like the host of stars hidden in the Milky Way, only visible with a spyglass, many minibeasts were so miniscule that they could only be appreciated through a microscope. Hardly part of God's plan. Other creatures hit even closer to home. The most arboreal of the great apes was the *orang hutan*, as it was known in its native Indonesia, where *orang* means "person" and *hutan* means "forest." The "person of the forest" was a canny creature indeed, and strikingly human. Curiously, neither did he, nor many of the other exotic flora and fauna, seem to have made it onto the guest list for the last voyage of Noah's Ark.

Though Darwin left England a creationist, his head must have been abuzz with such debate. On his sea voyages, he devotedly pored over Lyell's book, but it was not a comfortable journey for the young naturalist. Sitting in his bunk as the waves whorled and snarled against the rounded hull of the ship, Darwin soon suffered from the first bouts of seasickness that would afflict him over the next half decade of the journey.

BEAGLING

Launched from London's Woolwich Dockyard almost a dozen years before, the HMS *Beagle* was ten-gun sloop-of-war brig with two masts, about 27 meters long and 7 meters wide. As a ship-of-war, it had a full complement of 120 crew. But this was reduced to sixty-five, plus nine supernumeraries, for its second voyage, which with Darwin on board would eventually make the *Beagle* one of the most famous ships in history.

But Darwin was about to embark on another voyage, a personal journey with universal themes. This journey would force both him and his contemporaries to confront some very trying realizations about the Earth's natural history. All science needs a paradigm, a framework in which to develop hypotheses and test them. Darwin's paradigm was an account of evolution developed directly from Lyell's work. The foundation stone was gradual change: that the Earth is ancient and continues to change today, just as it did in the past.

Indeed, Lyell had also spent much of his life scouring the globe, looking for evidence to support a working hypothesis. In South America he had calculated that an earthquake could throw up coastal mountains by as much as a meter. Recurring quakes would have a cumulative effect: "A repetition of two thousand shocks, of equal violence, might produce a mountain chain one hundred miles long and six thousand feet high."[4]

More telling was Lyell's observations on the bioenvironment. The discovery of warm-water seashells in northern Italy, and of woolly mammoths frozen in the Siberian tundra, led Lyell to the conclusion that the climate in Europe was at one time "*sufficiently mild to afford food for numerous herds of elephants and rhinoceroses,* of species distinct from those now living" (the emphasis is Lyell's). And as Darwin noted, "The great merit of the Principles was that it altered the whole tone of one's mind, and, therefore, that when seeing a thing never seen by Lyell, one yet saw it partially through his eyes."[5]

Despite the caveat from John Stevens Henslow, his mentor at Cambridge, Darwin derived the same radical conclusions as Lyell; "I feel as if my books came half out of Sir Charles Lyell's brain,"[6] he would later quip. The expedition of the HMS *Beagle* provided Darwin with a rare, if not once in a lifetime, opportunity to study the planet at firsthand.

Darwin was confronted with a world rich in diversity. The HMS *Beagle* sailed across the Atlantic, around the southern coasts of South America, returning via Tahiti and Australia, having circumnavigated the Earth. While the expedition was originally planned to last two years, it lasted almost five. Darwin's explorations were carried out on foot, and on horseback, in deep caverns and on soaring mountain summits, across ice packs and searing sands.

He seems to have been well suited to the task of observing the flora and fauna of the world at close quarters.

> Nothing escaped him. No object in nature, whether flower or bird, or insect of any kind, could avoid his loving recognition. He knew about them all ... could give you endless information ... in a manner so full of point and pith and living interest, and so full of charm, that you could not but be supremely delighted, nor fail to feel ... that you were enjoying a vast intellectual treat to be never forgotten.[7]

According to Darwin, he "worked on true Baconian principles, and without any theory collected facts on a wholesale scale."[8] Francis Bacon had been a contemporary of Galileo's, an English spin doctor of the New Philosophy. He had believed in an inductive method. Nature could be pinned down, Bacon believed, by amassing materials, and drawing conclusions from the sheer weight of evidence. The mass of facts would ultimately lead to the truth. Darwin took the teaching to heart; he collected so many samples, his shipmates teased he was out to sink the *Beagle*.

At first, Darwin had no doubt in Scripture. Like many geologists and biologists of the early nineteenth century, he considered all species of life to have been simultaneously and individually created; he did "not then in the least doubt the strict and literal truth of every word in the Bible."[9] But his observations on the voyage, and the lucid and vivid writing of Lyell, began to change his mind.

In particular, Lyell was a dab hand at destroying the case of the catastrophists. The creed of Catastrophism held that the tortured and twisted state of rock could only be explained by the unique violence of the early Earth. But Lyell felt that "geologists have been ever prone to represent Nature as having been prodigal of violence and parsimonious of time."[10] It was far more likely, thought Lyell, that the state of the Earth was down to the ravages of time.

Lyell's example to Darwin also extended into biology itself: "The catastrophists relegate extinction to brief cataclysms, but if we then turn to the present state of the animate creation, and inquire whether it has now become fixed and stationary, we discover that, on the contrary, it is in a state of continual flux—that there are many causes in action which tend to the extinction of species, and which are conclusive against the doctrine of their unlimited durability."[11]

As to the nature of the fossil record, with its breakages, extinctions, and "missing links," Lyell maintained that catastrophe alone was not the explanation for such a state of affairs:

> Forests may be as dense and lofty as those of Brazil, and may swarm with quadrupeds, birds, and insects, yet at the end of ten thousand years one layer of black mould, a few inches thick, may be the sole representative of those myriads of trees, leaves, flowers, and fruits, those innumerable bones and skeletons of birds, quadrupeds, and reptiles, which tenanted the fertile region. Should this land be at length submerged, the waves of the sea may wash away in a few hours the scanty covering of mould.[12]

The message for Darwin was clear. If Lyell was right, then, "the causes which produced the former revolutions of the globe,"[13] continued today. And that had profound consequences in time. The age of the Earth must be measured in millions not thousands of years.

But there was another crucial decision for Darwin to make. As well as the contrast between a young and an old Earth, there was a choice between the Catastrophism of his creationist upbringing, and the Uniformitarianism of his naturally enquiring mind. At heart, this was a choice between idealism and materialism, between a closed philosophy of science, which ignored the new evidence of the fossil record for the sake of Scripture, and an open philosophy, which boldly went where angels feared to tread, following the clues to an unknown conclusion.

ZOÖNOMIA AND EVOLUTION

Darwin returned home a changed man. His journey with the HMS *Beagle* had provided plentiful firsthand experience that the Earth was still engaged in continual change. But his thoughts went one step further. He began to speculate as to whether species might also morph and whether that morphing might result in new species being created.

Luckily, Darwin was gifted with a familial vision that proved hard to live down. His grandfather Erasmus Darwin was an ingenious mechanic, inventing a speaking machine, a mechanical ferry and a rocket motor, long before the dreams of Russian rocket pioneer Konstantin Tsiolkovsky.[14] Erasmus was also a provocative evolutionist. As a boy Charles had poured over Erasmus's mighty work on evolution, *Zoönomia*, published in two volumes in 1794 and 1796.

It was replete with hearty exclamations that life had evolved from a single ancestor.

Erasmus Darwin was a celebrated communicator of science. Romantic poet Samuel Taylor Coleridge declared him "the first literary character of Europe, and the most original-minded man."[15] One of Erasmus's poems on evolution enjoys a science-fictional vision. It foresees, with unerring accuracy, a future of colossal skyscraper cities, overpopulation, convoys of nuclear submarines, and the advent of the car.[16]

Charles did not inherit his grandfather's boldness of spirit. Much has been made by biographers of the young Darwin's supposedly wasteful youth. His early years had been characterized by activities such as gambling, drinking, and the curiously prescient habit of obsessively collecting twigs and stones. Indeed, his father, Robert Darwin had complained that Charles seemed to "care for nothing but shooting, dogs, and rat-catching, and you will be a disgrace to yourself and all your family."[17]

Darwin had further frustrated his father by dropping out of Edinburgh medical school, having neglected his medical studies to investigate marine invertebrates. Robert Darwin was a wealthy society doctor and financier, who had studied medicine at the University of Leiden and taken his MD at Edinburgh, when he was only twenty years of age.

He had cautioned young Charles against voyaging on the *Beagle*, but eventually yielded. An imposing man, standing at six feet two and having reportedly stopped weighing himself when he was 24 stone, when Charles returned home after his five year journey, Darwin senior's first exclamation was, "Why, the shape of his head is quite altered."[18] Like many in the Victorian age, the elder Darwin was a keen phrenologist, believing that personality traits could be divined by "reading" bumps and fissures in the skull.

But pseudoscience aside, the interior landscape of Darwin's brain had indeed undergone a deep change. It could not have been a realization unique to Darwin. As has been shown, the evidence for evolution had been accruing throughout the eighteenth and nineteenth centuries, based on the experiences of the age of canal and railway building. It made any other explanation very hard to believe. And with this industrialization came the reality: animals and plants were once very different from what they are now.

Nevertheless, it had taken generations of geologists and biologists before Darwin to come to this conclusion. Years of painstaking and obscure work before the world would sit up and begin to embrace the idea of organic evolution. The voice of Darwin's grandfather Erasmus was one of the most prominent of those advocating the cause of change:

> Perhaps millions of ages before the commencement of the history of mankind, would it be too bold to imagine, that all warm-blooded animals have arisen from one living filament, which The Great First Cause endued with animality? ... What a magnificent idea of the infinite power of The Great Architect! The Cause of Causes! Parent of Parents![19]

Another great influence was the evolutionary view of the French naturalist Jean-Baptiste de Lamarck. According to Lamarck, traits acquired by individuals through experience in their lifetime could be passed down to their offspring. In Lamarck's world, thoroughbreds who excelled at racing bestowed their swiftness on their brood. Giraffes who extended their necks that extra inch to reach the sweeter delicacies on the taller trees conferred on the next generation the same long-necked advantage.

There was also a strong anthropological strain to Lamarck's theory. It was replete with moral insinuations. Victorians who led clean, conscientious, and vice-free lives bore children who were genetically inclined the same way, though just yet genes did not come into it, of course. But the one way in which all earlier theories of evolution floundered was on the question of how new species arose. Sure, the likes of Lamarck might try explaining away well-evolved creatures such as crocodiles and sharks, but not the origin of such species. As to the fossil record and its extinctions, the silence was deafening.

THE ORIGIN OF SPECIES

The great discovery of the nineteenth century, in which Darwin is deemed to have played a part, was not just to argue that life had evolved. That much had been argued before. Rather, the true innovation of the age was to identify the evolutionary mechanism by which new species came to be. After all, that is why, for his contribution to the debate, Darwin named his book, *The Origin of Species*. "Evolution" was not a word he liked to use.

Like Galileo's *The Starry Messenger*, Darwin's book was to send out great reverberations through the learned world, and far more importantly beyond academia and into the culture. For Darwin, like Galileo, had made his book readable for the layman, further adding to the widespread interest on publication. But, unlike Galileo's shot across the bows gathered together in only a few dozen pages, Darwin's book is a thorough work, a magnum opus that is patiently developed over hundreds of pages. Indeed, even the original title was *On the Origin of Species by Means of Natural Selection, or the Preservation of Favoured Races in the Struggle for Life.*

The case outlined in *The Origin of Species* can be distilled down into three component concepts.

The first notion, and one inextricably linked with industrialization in Victorian Britain, was "variation." Variation is based on the observation that each and every individual of any particular species is different. Victorians understood this with consummate ease. The process of industrialization brought laboring people in from the country along with a fascination for animal and plant breeding. Darwin's father was a pigeon fancier, and his father-in-law, the celebrated English potter Josiah Wedgwood, a renowned sheep breeder. From such animal husbandry an all-important lesson was to be learned: how one might promote or inhibit the subtle characteristics of a breed.

The second concept was "multiplicity." Here the idea is based on another observation of living creatures; that they tend to make more offspring and have bigger broods than the environment can necessarily maintain. Only a decade before, English poet Alfred Lord Tennyson had famously written about, "Nature, red in tooth and claw." Indeed, it is a cruel world. One in which only a small fraction of the baboons and badgers and bugs that are brought into existence actually survive, or manage to evade predators long enough to mate.

On this question of "multiplicity," the conventional pro-Darwin narrative is that English political economist and demographer, Thomas Malthus, had been very influential. Darwin had read Malthus's *An Essay on the Principle of Population*, so the story goes, some forty years after Malthus had first anonymously published the book. In his work, Malthus had suggested that most species reproduce geometrically. The environment, however, can maintain little better than a linear increase in species populations. And so apparently Malthus, who had been partly inspired by reading Darwin's grandfather Erasmus, was in turn one of the key influences on Darwin himself.

"One may say," Darwin wrote, that "there is a force like a hundred thousand wedges trying to force every kind of adapted structure into the gaps in the economy of nature, or rather forming gaps by thrusting out weaker ones."[20] If Darwin is to be credited for a contribution to the development of the hypothesis of evolution, then Malthus is meant be one of his sources. Malthus's work focused on a kind of "natural selection," a recipe based on the boundless productivity of nature, coupled with the limited resources on hand to maintain them.

Here, anyhow, was a natural and global mechanism. One that worked continually to snuff out most variations, while conserving those few carried by individuals who had won the struggle to survive and mate. Here was "natural selection," the third component concept covered in Darwin's famous book. The individual differences between members of a species, coupled with the environmental forces highlighted by those like Malthus, shape the likelihood that a particular individual will last long enough to pass its characteristics on to posterity.

Consider moths. In snowy weather, white moths do better, since their color acts as camouflage, protecting them from predators. Russet moths fare better in the fall, where they are easily lost against the autumn trees. Indeed, a remarkable illustration of the results of such adaptive color change happened in Victorian Britain.

Naturalists had noted in the eighteenth century that all of the British peppered moths in the district of Manchester were pale in color. But in 1849 a solitary black moth was found in the area. The industrial impact of the infamous "dark Satanic mills" of William Blake's poem had blackened tree trunks and wreaked havoc on the white moth population in that part of Lancashire. Their original camouflage had been stripped of its advantage, inadvertently granting unsolicited gain upon the few black moths in the area. Following a decision by the local municipal authority to try curbing industrial blight, the soot was slowly washed away (Manchester is also renowned for its rain), and the pale peppered moth population bounced back.

This rather charming story of the moth population suits us well. It illustrates how the "fittest" survive, to use the phrase coined by the English political theorist Herbert Spencer. But what makes them "fit" is not some kind of innate superiority to others. Rather, it is an unsolicited advantage, the result of the pure serendipity of freak change; they just happen to better "fit" their environment. Should the scene change in time, those previously well adapted may swiftly discover they no longer fit. Soon, other freaks may inherit the Earth, as it were.

Darwin's *Origin of Species* identifies natural selection as the mechanism that creates new species. Twenty-three centuries after the Ionian philosopher Heraclitus suggested everything was in a state of flux, Victorian science saw a world in a state of constant change. The emerging science of biology had identified two things. First, that nature favored variety. A population of pale moths would profit from the odd dark moth, for who knows what the future may bring. Second, nature also prefers geographical spread. For the further afield a species is scattered, the less tied their prospects to a single setting.

Accordingly, the measure of individual variations in a particular species is apt to grow over time. Eventually, some clusters will have become so diverse from other clusters that mating and reproduction are no longer viable. Such a coupling would not produce a fecund litter. A new species would emerge. Here is Darwin's take on the process:

> During the modification of the descendants of any one species, and during the incessant struggle of all species to increase in numbers, the more diversified the descendants become, the better will be their chance of success in the battle for life. Thus the small differences distinguishing varieties of the same species, steadily tend to increase, till they equal the greater differences between species.[21]

Like Galileo, Darwin was a devout man, at first anyhow. But their observations and explorations in the natural world led to an inescapable conclusion. The universe was not as Scripture said. When Galileo became an advocate for the Copernican system, he tried changing the Church's interpretation of some passages of Scripture. Perhaps the most famous example is Joshua's command that the Sun stand still. If Galileo was right, Joshua's plea was pointless; it was the Earth that was in orbit, not the Sun.

Unlike Galileo, however, Darwin at first tried accommodating Scripture. He noted the resemblance between the theory of evolution and the biblical notion of the Tree of Life. His approach was one of gradual conciliation. But rather than his Tree image being static, as with the creationist case, the new vision was one of change, albeit gradual.

> The green and budding twigs may represent existing species; and those produced during former years may represent the long succession of extinct species … The Tree of Life … fills with its dead and broken branches the crust of the earth, and covers the surface with its ever-branching and beautiful ramifications.[22]

His appeasing ways often fell on deaf ears. Biblical literalists grumbled that natural selection was cold and mechanical. In Darwin's mind, however, the theory identified the lifeblood of the natural world.

> When we no longer look at an organic being as a savage looks at a ship, as something wholly beyond his comprehension; when we regard every production of nature as one which has had a long history; when we

contemplate every complex structure and instinct as the summing up of many contrivances, each useful to the possessor, in the same way as any great mechanical invention is the summing up of the labour, the experience, the reason, and even the blunders of numerous workmen; when we thus view each organic being, how far more interesting—I speak from experience—does the study of natural history become![23]

And to this mollifying approach to clerical prejudice, Darwin further developed the case of the evolutionist.

There is a grandeur in this view of life, with its several powers, having been originally breathed by the Creator into a few forms or into one; and that, whilst this planet has gone cycling on according to the fixed law of gravity, from so simple a beginning endless forms most beautiful and most wonderful have been, and are being evolved.[24]

According to convention, Darwin had amassed the bare bones of his theory as early as 1839, and that by 1844 he had developed it further, in the form of a 230 page treatise. And yet he had held back. Darwin had waited another fifteen years before publishing his magnum opus. Why the dubious delay?

The jury is still out. One theory puts the delay down to illness. From his marriage in 1839 and possibly way before, Darwin suffered severe headaches, vomiting, and uncomfortable palpitations of the heart. According to his son Francis, Darwin's life "was one long struggle against the weariness and strain of sickness."[25] Cures were certainly sought. But when diagnosis is doubtful, treatment can be as much a game of chance as evolution itself.

The finest physicians in England were consulted. And it seems some of the curative crazes of the Victorian age were tried and tested on the famous biologist. Hypnosis was one, hydrotherapy another. Darwin was forced to spend miserable winter days swathed in a cold, wet sheet. But his illness was never decisively diagnosed. One guess was Chagas' disease. This is a tropical parasitic disease caused by the blood-sucking assassin bugs of the pampas, which Darwin may have contracted on March 26, 1835, while voyaging on the *Beagle*.

Another theory is stress. The source of such stress, and its possible causes, has attracted much debate. Some scholars have focused on Darwin's submissive relationship with his father, or his passive dependence on his wife. Others have concentrated on the psychosomatic results of Darwin's personal journey, from potential priest to Devil's chaplain.

Indeed, the words are Darwin's own. "What a book a Devil's chaplain might write on the clumsy, wasteful, blundering low and horridly cruel works of nature!" he confessed in his private note-book. Here for many is the main reason for Darwin's reluctance to publish—he was terrified of the very idea of evolution. Terrified of the blow to Christianity, and terrified of what such a theory might do to the dignity of man. Wedded to respectability and social order, Darwin feared that the theory would inspire "atheistic agitators and social revolutionaries."[26]

For evolution struck at the heart of humanity itself. Newton's system of the world had essentially reestablished the integrity of design, which had been shattered by Galileo. The Christian picture of creation had stayed more or less untouched. Man was still made in the image of God. Darwin suspected that after evolution, the book of Genesis would lay in shreds as a literal history.

Resistance from the Church promised to be far more than futile. Like Galileo before him, Darwin would not have to trouble his imagi-nation too much to anticipate the antipathy that would come his way, once his book was published. Galileo had provided evidence that the so-called central Earth was adrift in space, in orbit about the Sun. An infinite universe was only a conceptual quantum leap away.

Darwin was about to publish on a theory that would be as damag-ing a break to religious tradition in the animate world as Galileo's had been in the inanimate. Copernicus had kept his cosmology a secret. He well understood how the position of the Earth was vital to the Christian drama of life and death in the Middle Ages, how religion was adjusted to the great plan of the universe, and how that plan would be shattered by the shifting the Earth from its "natural" place at the "corrupt" center of the universe. For fifteen years Copernicus failed to let loose his cosmology. In the end he relented, and *De Revo-lutionibus* was published on his deathbed, still a sobering decision for a devout man.

Copernicus set in train a revolution, the birth of a new physics. The theory of evolution would assert that all life on Earth had a common ancestor, that men and animals are kin, and that chance mutations, not divine majesty, drove history. In time, the theory of evolution was to become the foundation of biology, a unifying science of the diversity of life.

It was murder, according to Darwin. In an admission to his friend the botanist Joseph Hooker, he compared the public promotion of evolution to confessing murder. Indeed, support of the theory would later be referred to as "the murder of Adam." Galileo had faced the wrath of the Holy Roman Empire for promoting Copernicus. But for endorsing evolution, Darwin need not have looked beyond Britain to envision what might lie in store from religious quarters. Besides, like Copernicus and Galileo, Darwin would be defending a theory he knew to be incomplete.

And one that he may very well have stolen from Alfred Russell Wallace.

PART IV

THE AFTERMATH: WORLDS TURNED UPSIDE DOWN

CHAPTER 8

THE "GALILEO" AFTERMATH

It is early morning, sometime in September 1638. A yawning Englishman is making his way from the center of the city to a villa, south of the river. As this poet wends his way through winding streets, you too can hear the hushed city breathing, a city esteemed for the elegance of its dialect, its genius, and its taste.

The poet's long walk begins at the Santa Maria Novella, where the world's most infamous scholar was not so long ago attacked from the pulpit for his views. Listen. The bell tower of Giotto sounds in the Duomo as the poet wanders through the streets and alleys of the city that helped nurture the dawn of modern science.

Slowly, he makes his way through the still sleeping streets, but your eyes are unclosed to see the black and misty river, snaking its way to the sea. And you alone watch as the poet, the river on his right, makes his way past the Santa Croce where, a century hence, the scholar's tomb will sit among a wealth of art and architecture.

Listen. As the poet knocks on the door of the great scholar, now old and blind, confined under house arrest, by order of the Inquisition.

The poet is a man of a mere thirty years. His own blindness, his own arrest, and his own cosmological epic, all lie before him. The scholar is in his seventy-fourth year on this Earth, one which revolves around the Sun, by his own affirmation, a view which five years before he was forced to recant, under threat of torture.

The poet is John Milton. The home on which he calls is the Florentian prison for the last half decade of Galileo, incarcerated for his *Dialogue Concerning the Two Chief World Systems*, among other things.

A fervent republican in an age of absolute monarchies, Milton's encounter with Galileo left a deep imprint on the man who was to become the poet of the English Revolution. It crept into his *Paradise Lost*, where the shield of Satan looks like the Moon, seen through Galileo's spyglass. And in *Areopagitica*, Milton's great defense of free speech, he evokes his visit to Galileo in Florence, warning that England too will buckle under inquisitorial forces if it bows to censorship, "an undeserved thraldom upon learning."[1]

There is something so resonant about imagining these two figures inhabiting the same age. The new world system of the old astronomer's empirical universe meets its match in Milton's complex aura, the sense of new and old worlds in collision, the upending of cosmologies and kings.

Copernican astronomy had gifted the radicals of the English Revolution a view of the universe as a whole, science and society as one. The distinction between heavenly and sublunary spheres had ended. The radicals aimed at bringing this change down to Earth by ending the distinction between specialists and laymen. They wanted an end to the dominance of Latin, Hebrew, and Greek. They wanted to rid the universities of their scholastic theologians, and to drive kings into oblivion.

The last thing the radicals wanted was for the emerging science to be handed over to a new set of mumbo-jumbo men.

Mystic visionaries such as John Dee, Giordano Bruno, and the English physician Robert Fludd had hoped to comprehend the whole cosmos, in all its aspects. The radicals wanted science, philosophy, and politics taught in every parish, by an elected nonspecialist. Along with the radical scientists, they wanted science to be rationally applied to the problems of human life.

Indeed, they wanted democratization of all things, a Commonwealth of knowledge in which the yawning gap between the useless specialized scholar and the ill-educated practical men would be slammed shut. In their vision of society, the two cultures would have been one.

The enemy was monopoly. For centuries, knowledge of the soul had been shut up in the Latin Bible, which only black-coated ministers had to interpret. The administration of justice was monopolized by lawyers and judges, medicine by the College of Physicians. Was science now to be shut up in its own jargon, which only new specialists were to interpret?

Ironically, their vision of the democratization of science, and the widespread dissemination of knowledge, came during trying times,

times when those in positions of power made moves to secure special-
ization in the hands of the few. The last of the polymaths were dying
out, just when the radicals called for a minipolymath in every parish.

And over in Italy, Galileo did little for their cause. A drama was
about to unfold that sold out science to the Church, and handed the
revolution over to the forces of reaction.

THROUGH A GLASS, DARKLY

Not everyone believed Galileo's discoveries with the "optick tube."

Copernicus had been a quiet, unassuming man all his life. No one
who met the captivating Kepler could seriously dislike him. Galileo
was another matter. Portraits show a red-haired, thickset man with
rough features and an arrogant stare. Galileo held radical leanings and
a contempt for authority that would soon lead to infamous trouble.
His exceptional talent for antagonizing others led to skepticism in the
small academic world of his own country.

On the evenings of April 24 and 25, 1610, a memorable party
was held to celebrate Galileo's recent discoveries. The great man
was invited to demonstrate the Jupiter moons in the spyglass. Not
one among the eminent guests was convinced of their existence. The
crude nature of the mysterious gadget didn't help, but many were
blinded by prejudice. They even refused to look down the tube.

Over the next year or so, Galileo's celebrity was in the ascendant.
The Starry Messenger saw a number of reprints, and more and more
astronomers were turning their telescopes to the sky. But once more
due to scopes vastly poorer than the Tuscan's own design, they saw
little. There was no sign of the moons of Jupiter. Even the mountains
of the Moon were indistinguishable. And again there were others, a
gaggle of philosophers throughout Europe, who just plain declined
to look through the eyepiece.

There are none so blind as those who will not see.

So faulty scopes and narrow minds did not help the Copernican cause. Neither did Galileo himself. Galileo's arrogance aroused
uproar. A public controversy followed similar to the UFO debacle
three hundred years later. Claims of optical illusions, haloes, and the
unreliability of inexpert witness. As Galileo's discoveries had become
the talk of the world's poets and philosophers, scholars in his own
land were skeptical or downright hostile.

In Bologna, a young upstart of an astronomer by the name of
Martin Horky added his name to the list of skeptics. His *A Very Brief
Pilgrimage Against the Starry Messenger* was in fact a very amateur

broadside on Galileo's discoveries. It seems the aim was merely to become famous for attacking someone famous. The plan backfired. Kepler, Horky's former teacher, wrote to Galileo greatly regretting the act of "this scum of a fellow."[2]

Another attack, that of Francesco Sizzi, was grounded in theology. His *Dianoia Astronomica, Optica, Physica* firstly pointed out that there could be no moons of Jupiter, since they were not so described in Holy Scripture. It went on:

> Just as in the microcosm there are seven "windows" in the head (two nostrils, two eyes, two ears, and a mouth), so in the macrocosm God has placed two beneficent stars (Jupiter, Venus), two maleficent stars (Mars, Saturn), two luminaries (Sun and Moon), and one indifferent star (Mercury). The seven days of the week follow from these … Moreover, the satellites are invisible to the naked eye and therefore can have no influence on the Earth, and therefore would be useless, and therefore do not exist.[3]

It was hardly a sound scientific case, and one that was easily dealt with by Galileo himself. Sadly, Sizzi's futile treatise was dedicated to a member of the local and noble Medici family, of whom more will be discussed later. As a result, Sizzi's nonsense reached a far wider audience than it would otherwise have done.

From the outset it is interesting to note that Galileo's foes were of many different hues. Some were the conservative Aristotelians at the universities. Others were religious fanatics, made more dangerous in an age controlled by such extremists. Such a man was the philosopher Ludovico delle Colombe, a man who despised Galileo, not just for what he had found, but also for the way in which Galileo ridiculed him, just after the publication of *The Starry Messenger*.

Colombe claimed that Galileo was incompetent, and that his sightings were either bogus, due to imperfections on the spyglass lenses, or misconstrued by an overly zealous mind. Colombe clung on to the comfort of Scripture in the face of the new discoveries of science: "He who would render false all the belief of mathematics, philosophy and theology, who dares to demonstrate against all received wisdom and communication, that is a person who would put an end to Holy Scripture."[4]

KEPLER TO THE RESCUE

The strongest voice raised in Galileo's defense belonged to Kepler, whose position as first astronomer of Europe was by now beyond dispute.

Johannes Kepler had apparently been conceived at 4:37 a.m. on May 16, 1571, and was born prematurely on December 27 at 14:30, after a pregnancy lasting 224 days, 9 hours, and 53 minutes. Such unerring accuracy, culled as it is from his own astrological charts, paints an immediate portrait of Kepler as a man of great contrasts and contradictions, typical of an age of transition.

He was said to have stemmed from a noble family, but by the time of his birth the line had fallen into degenerates and psychopaths. Kepler's mother was raised by an aunt who was later burned alive as a witch. His mercenary father narrowly escaped the gallows. An altogether curious pedigree, but perhaps telling for a man who was to become the most brilliant and erratic speculator of the Scientific Revolution.

Kepler had got his first job in 1593, as a professor of Mathematics in Gratz, Austria. It brought mixed rewards. Kepler was a brilliant scholar, but a rather poor teacher. Whenever he got excited, and he was almost always in this state, he "burst into speech without time to weigh whether he was saying the right thing."[5] His digressions constantly led him to think of "new words and new subjects, new ways of expressing or proving his point, or even altering the plan of his lecture."[6] Hardly any wonder that in his first year of teaching he had only a small handful of students, and in his second, none whatsoever. He was the very picture of an absent-minded unintelligible professor, delivering garbled lectures before an empty classroom.

Whereas Galileo was wholly and frighteningly modern, Kepler never severed himself from the mystical Middle Ages. Unlike Galileo, who was devoid of any spiritual leanings, Kepler was struck by the magical implications of the Sun-centred universe.

Kepler was fanatical about discovery. The spirit of scientific inquiry seemed to well up within him. His books on astronomy attempted to lay bare the ultimate secrets of the cosmos. Yet they were a hotchpotch of geometry, music, astrology, astronomy, and the occult. Indeed, Kepler's famous three laws of planetary motion are buried deep within such a work of lavish fantasy.

The stunning intelligence of Galileo's *Starry Messenger* had arrived at Kepler's door in the hands of Wackher von Wackenfels in March of 1610. By the beginning of April, Kepler had his own copy to study from Galileo himself, along with a request for Kepler's opinion. Though Kepler had no spyglass of his own, and no means of verifying Galileo's findings, he nevertheless gave Galileo his wholehearted support.

It came in the form of Kepler's pamphlet, *Conversation with the Starry Messenger*, published in April 1610. The booklet enthusiastically endorsed Galileo's observations: "Perhaps I shall be considered reckless because I accept your claims as true without being able to add my own observations. But how could I distrust a reliable mathematician whose art of language alone demonstrates the straightness of his judgement?"[7]

And it presented an array of speculations about the meaning and implications of Galileo's discoveries, for astronomy as well as cosmology and astrology: "In the battle against the grumpy reactionaries, who reject everything that departs from the beaten track of Aristotle as a desecration,"[8] Galileo's work offers, "a highly important and wonderful show to astronomers and philosophers, that it invites all friends of true philosophy to contemplate matters of the highest import."[9]

The support of Kepler was of immense significance. Later that year, he published his own telescopic observations of Jupiter's moons in *Narratio de Jovis Satellitibus*, providing further support for Galileo. Kepler was the Imperial Mathematicus to the Holy Roman Emperor, Rudolf II, the elected monarch ruling over the entire Holy Roman Empire, a Central European state in existence throughout the Middle Ages, and into the Early Modern period.

THE TELESCOPE AND THE ALIEN

There is another great significance of *The Starry Messenger*: it inspired in Kepler early and uncannily prophetic ideas of alien life.

Though Kepler's *Conversation with the Starry Messenger* is a rather rambling and hurried response, it contains some fascinating insights into his ideas on Copernicanism, life on other worlds, and space travel.

> There will certainly be no lack of human pioneers when we have mastered the art of flight. Who would have thought that navigation across the vast ocean is less dangerous and quieter than in the narrow, threatening gulfs of the Adriatic, or the Baltic, or the British straits? Let us create vessels and sails adjusted to the heavenly ether, and there will be plenty of people unafraid of the empty wastes. In the meantime, we shall prepare, for the brave sky-travellers, maps of the celestial bodies—I shall do it for the Moon, you Galileo, for Jupiter.[10]

For Kepler had his own startling story to tell. Just one year before Galileo's earth-shattering discovery, he had published in draft manuscript the first ever work of science fiction in *Somnium*. Kepler's tale was a space voyage of discovery in the new physics that invented the alien, and anticipated the universe soon to be unveiled by the telescope.

Somnium is a truly extraordinary work. Its theme is the new universe through Galileo's spyglass. There is seldom a passage in Kepler's twenty solid volumes of writings that is not alive and kicking. So it is with *Somnium*. As a fictional travelogue it follows on from ancient Greek stories, such as Plutarch's *The Face on the Moon* and Lucian's facetiously titled satirical work, *A True Story*. In all other ways it signals a sharp break with classical tradition.

The change in the economy had other effects on the material conditions of Galileo and Kepler's time, compared to that of the ancient Greeks. The development of the printing press had transformed the spread of knowledge. There was an increasingly diverse readership and publication rates rocketed. In England, for instance, only eighty books were published per year in the 1540s, but by the 1640s this figure had exploded to four thousand. No doubt Kepler was aware of the growing use of print as an ideological battleground. His science fiction tale suited the traditional oral culture. By the 1620s the telling of sensationalist stories with a moral purpose became a very popular form.

The story of *Somnium* is an extrapolated voyage of discovery. Copernicus had shifted the center of the universe to the Sun. Galileo had provided the evidence with the telescope. Kepler's aim was to explore this alien panorama from the alternative standpoint of the Moon. He wanted to describe what the new astronomy would be like from the perspective of another planet.

In this way the feasibility of a non-geocentric system could be explored. Novel observations of the heavens, and the Earth itself, could be made. On his journey, the "spirit of the Moon" transports Kepler's hero Duracotus to a lunar landscape. Once there, he gazes down upon the Earth, looking at its geography, its motion through space, also exploring the surface of the Moon and its alien inhabitants.

Somnium is a watershed. It marks the end of the old era and heralds the arrival of the new science. Its hypothesis on extraterrestrial life had an estranging effect, revealing the world in a new light. Kepler's book is an imaginative tour de force. It was the first space fiction of the age. The alien voyages quickly evolved as a potent motif for exploring the insignificance of man.

Somnium's influence was huge, inspiring other interplanetary journeys such as those of John Wilkins, Henry More, H. G. Wells, and Arthur C. Clarke. Kepler was a pioneer of the new vision of deep space as the home of a plurality of inhabited worlds.[11] There is no greater testament to the power and imaginative sway of the telescope than this. Despite the incredible odds against detecting life beyond our planet, billions have been spent in the twentieth century on sober scientific projects in the search for alien intelligence. That search started with the telescope.

The Scientific Revolution begins with the discovery of these other worlds, symbolized by the names of Kepler and Galileo. They produced a map of the knowable, just as the unknown was at the point of becoming known. It was Galileo's use of the telescope that started to shatter Aristotle's crystalline universe. But Kepler's fiction went one step further. *Somnium* implied that creatures may well be dwelling there. It was a vital new piece of evidence in the debate on the existence of extraterrestrial life.

And it is the first time the Moon becomes a real object for us. At that same instant we feel a sense of wonder, or estrangement, from this new reality. Estrangement implies a state of imperfect knowledge. It is the result of coming to understand what is just within our mental horizons.[12]

Galileo had to assume that shadows on the Moon have similar causes to shadows on Earth in order to understand the Moon's difference from Earth. Yet so great an astronomer as Kepler evidently needed to believe in extraterrestrials in order to render Galileo's discovery thinkable. Kepler selected the framework of *Somnium* to pass off his Copernican essay as a dream. In this way he was able to subvert the scorn of the Aristotelians by concealing his radical work in the guise of classical mythology.

Kepler realized that to understand the Moon it was not enough to put one's observations into words. The words themselves had to be transformed by a new sort of fiction. That's why there is something revolutionary about *Somnium* in the history of science. Throwing words at the Moon, as it were, has a dialectic effect—the words come back to us changed. By imagining strange worlds, we come to see our own conditions of life in a new perspective.

A SHORT EXCURSION IN CRYPTOLOGY

Galileo was quick to reap the benefits of Kepler's backing. It was exactly what he needed to hear at that moment. The influence of Kepler's authority as a philosopher played a crucial part in turning the tide, not just for Copernicanism, but also for Galileo's career.

The Tuscan astronomer was keen to leave Padua behind, to be appointed court mathematician to Cosimo de Medici, the grand duke of Tuscany, in whose name Galileo had dubbed the Jovian moons "the Medicean stars." In his letter of application to the duke's secretary of state, Galileo gave prominent place to Kepler's weighty support:

> Your Excellency, and their Highnesses through you, should know that I have received a letter, or rather an eight-page treatise, from the Imperial Mathematician, written in approbation of every detail contained in my book without the slightest doubt or contradiction of anything. And you may believe that this is the way leading men of letters in Italy would have spoken from the beginning if I had been in Germany or somewhere far away.[13]

The advice Galileo received from the Tuscan ambassador at the Imperial Court was simple: send a telescope to Kepler. Given that Kepler had lent his support on trust, the least Galileo could do was provide the means by which more robust and evidentiary backing could be offered. Sadly, Galileo did not heed the call. Instead, he gifted the telescopes from his workshop to divers and noble patrons.

Nonetheless, one of the most fascinating episodes in the history of science communication was about to begin. A series of coded messages of discovery passed between Galileo and Kepler. The fascination is not merely in the content of the communication itself. The aim was apparent. Galileo's latest telescopic findings were sent as anagrammatic messages, coded sequences designed to prevent any pretender from stealing his claims of priority.

The fascination is also in the complementary styles of Kepler and Galileo. The actual content of Galileo's coded messages is stolid and matter of fact, mostly devoid of mystical leanings. The attempted solutions of the mercurial Kepler are flowery and fantastic. The different styles speak strongly of the contrasting characters of their authors.

It all started with Galileo sending a simple cipher to the Tuscan Ambassador in Prague. It read, "smaismrmilmepoetaleumibunenugttaurias."[14] The plan was to have the ambassador show the code to the keen Kepler, in the hope that he would grasp its meaning for everyone there. Kepler had a crack at it. His solution, in Latin, was, "Hail, burning twin, offspring of Mars,"[15] thinking that Galileo had found moons around Mars too.

Alas, Kepler was off the mark, but not by much. Galileo had indeed found more moons, or at least *thought* he had. His disclosed solution came some months later, "I have observed the highest planet [Saturn] in triplet form."[16] Galileo believed he had viewed two small moons on either side of Saturn, tucked tightly in to the planet.

In fact, and even though we can be sure he would have kicked himself had he realized, Galileo was the first to have seen Saturn's rings. It is just that his telescope was not powerful enough to resolve the detail of Saturn's beautiful ring system.

Indeed, it was another fifty years before Dutch physicist Christiaan Huygens became the first person to realize that Saturn was surrounded by a ring. Using a telescope that was far superior to those available to Galileo, Huygens observed that Saturn "is surrounded by a thin, flat, ring, nowhere touching, inclined to the ecliptic."[17]

But Saturn mystified Galileo. He described the planet has having "ears." And when in 1612 the plane of the rings was oriented directly at the Earth, the "ears" appeared to vanish. "Has Saturn swallowed his children?" wondered Galileo, only to be further confused when they reappeared again, a year later.[18]

Back in 1610, Kepler was also confused. Galileo's next coded message read, "Haec immature a me jam frustra leguntoroy" (These immature things I am searching for now in vain).[19] Once more Kepler tried several solutions. The most startling of which was, "Macula rufa in Jove est gyratur mathem, etc" (There is a red spot in Jupiter which rotates mathematically).[20]

Of course, there *is* such a spot on Jupiter, the Great Red Spot, large enough to engulf several Earths.

It is the vortex of a massive storm in the atmosphere of the planet, a storm that has raged and rotated persistently for 300 years. But it was not Kepler who found it. For many years the prevailing opinion was that the old "Permanent Spot" of Jupiter was discovered by English polymath Robert Hooke in 1664, though many affirm the primacy of Italian astronomer Giandomenico Cassini's discovery of 1665.[21]

Whatever the primacy, it is a striking curiosity. It is half a century before the spot is actually discovered, and over three hundred years before *Voyager* time-lapse movies dramatically revealed the spot's counterclockwise circulation in 1966.[22] And yet, Kepler seems to stumble upon this strange solution to Galileo's latest discovery by suggesting this red rotating spot on Jupiter. Even the color is correct.

The actual nature of Galileo's message was revealed a month later, "Cynthiae figures aemulatur amorum" (The mother of love [Venus] emulates the shapes of Cynthia [the Moon]).[23] It seems that Galileo had caught Kepler's mystic bug.

The Tuscan had made another shattering discovery: Venus in phase.

Like our own Moon, the planet Venus had been found by Galileo to show phases, waxing from slender sickle, through to full disc, waning down to a crescent once more. Such planetary phases are now well known, the phases depending on the relative position of the Sun, the planet, and the onlooker.

But the crucial thing is this: the onlooker sees the planet in phase, only when the planet is inferior. Superior planets, such as Mars, Jupiter, and Saturn, only ever appear as full or gibbous. In other words, only when the planet is closer to the Sun than the onlooker are the phases to be observed. Seen from space, the Earth itself would seem to go through such phases.

The model of Copernicus had predicted that such phases would be visible. The orbit of Venus around the Sun would cause an illuminated half to face the Earth when it was on the opposite side of the Sun, and to face away from the Earth when it was on the Earth-side of the Sun. In the Ptolemaic system, however, only crescent and new phases would be seen, since Venus was thought to remain between the Sun and Earth during its orbit around the Earth.

Galileo thought this was proof of the Copernican system, proof that Venus did indeed revolve around the Sun. The supporters of Aristotle held another view. They held to the new system of Tycho Brahe. No doubt Tycho had been a fine astronomer, his work on the nova of 1572 and the great comet of 1577 doing much damage to the cosmology of Aristotle.

But Tycho was no revolutionary. He was unable, or perhaps unwilling, to cast off the cosmology of Aristotle altogether. Though he lived in a liberal land with lenient attitudes to religious freedom, Tycho was devout. So, Tycho's system saw the Moon and the Sun revolve around the Earth, fixed at the static center of the universe. The other five planets, Mercury, Venus, Mars, Jupiter, and Saturn, revolved around the Sun, said Tycho.

The discovery of Venus in phase was arguably historically Galileo's most important use of the telescope. For the phases proved Venus orbited the Sun, and gave support to (but did not prove) the heliocentric model. But more importantly, since it proved the Ptolemaic geocentric model false, many seventeenth century astronomers now took flight. They abandoned the pure geocentrism of Ptolemy and joined the tribe of Tycho.

The Aristotelians claimed Tycho's system could also explain the phases of Venus, without the complication of Copernicus. It was a way of explaining the observations, they claimed, while at the same time appeasing the Church.

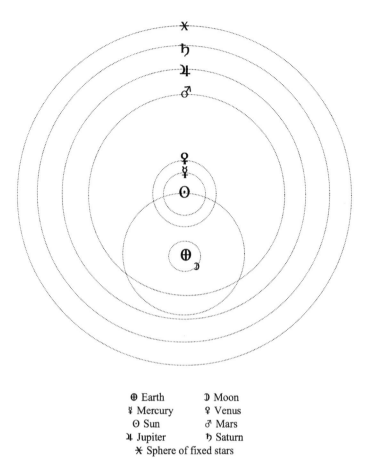

⊕ Earth	☽ Moon
☿ Mercury	♀ Venus
☉ Sun	♂ Mars
♃ Jupiter	♄ Saturn
✷ Sphere of fixed stars	

Figure 8.1 The Tychonic system: The Moon and Sun orbit the central Earth; the other planets (Mercury, Venus, Mars, Jupiter, and Saturn) orbit about the Sun. Pictured at the edge of this closed-in cosmos, is a sphere of fixed stars.

Galileo begged to differ, with good reason. For if the configuration of the solar system was as Tycho said, there would be turmoil. With the planets in orbit about the Sun as described by Tycho, Galileo reckoned that the tidal forces making this possible would surely also exert the same force on the Earth itself. With great upheaval, the Earth would alter its position and also fall in orbit about the central Sun.

The Tuscan had made the Dane's model seem like a logical absurdity, which it was. And if Tycho's model could be dismissed, and the conventional geocentric model was unable to explain the indisputable evidence of the telescope, that left the way open for the Copernican heliocentric model.

Galileo's discoveries with the telescope were becoming more Earth-shattering by the minute.

At last, Kepler was lent a telescope by one of the selected few noblemen to whom Galileo had gifted the gadget. Kepler was astounded to see the wonders at firsthand. Between August 3 and September 9 of 1610, he studied with fascination the findings catalogued by Galileo in his *Starry Messenger*, most notably the moons of Jupiter.

Shortly after, Galileo's findings of Jupiter were confirmed in Kepler's short pamphlet, *Observation-Report on Jupiter's Four Wandering Satellites*. Kepler's pamphlet was also hurriedly published in Florence. It was the first scientific and independent confirmation of Galileo's observation of the Jupiter moons.

It was also history's first use of the word "satellite," to mean a moon in orbit around a celestial body. No doubt the mystical Kepler still believed that Jupiter had some godlike cachet. In Roman mythology, Jupiter was king of the gods. And such powerful men were usually orbited by a knuckle of bodyguards, or *satellitem* in Latin. Rather fitting, then, that Kepler should use satellites for the Galilean moons in seemingly protective orbit about the planet of the same name.

The cryptic correspondence between Kepler and Galileo finally faded. In time, we shall see that Galileo's ignorance of Kepler's work was to cost him dearly with the Church. For, in ignorance of Kepler's laws of planets that moved in ellipses rather than epicycles, Galileo was forced into a flawed defense of circles as the only imaginable form of heavenly motion.

TOAST OF THE NOBLES

Back in 1610, *The Starry Messenger* became a triumph. The discovery of four new worlds around Jupiter was resounding. Its ringing tremors were felt throughout Italy, in the colleges and cloisters, in the piazzas and pulpits, and beyond, to the rest of Europe.

With a keen sense of the mercantile, Galileo had tried to sell the title of the Jovian "stars" in succession to the Duke of Florence (a Medici), to the King of France, and to the pope. Predictably, the cost of the heavenly honors proved too princely for all of them.

Galileo was flush with success. By September of 1610, he had taken up his new position as "Chief Mathematician and Philosopher" to the Medicis in Florence. Early the following year, he journeyed to

Rome. The trip was a great success. In a letter to Cosmo II, the Duke of Tuscany, in May 1611, Cardinal del Monte declared,

> Galileo has, during his stay in Rome, given great satisfaction, and I think he received as much, for he had the opportunity of showing his discoveries so well, that the learned and notable in this city found them no less true and well founded than astonishing. If we were still living under the ancient Republic of Rome, I verily believe that there would have been a column on the Capital, erected in Galileo's honour.[24]

Pope Paul V had received Galileo, extending him a warm audience. The chief astronomer at the Jesuit Roman College was none other than the venerable Father Clavius, lead author of the Gregorian Calendar reform. *The Starry Messenger* had at first mostly amused Clavius. He now gave it his wholehearted support.

On the same visit to Rome, and for the first time in history, the word "telescope" was used in relation to the spyglass.

Galileo had been attending a banquet in his honor, organized by the select Lincean Academy. Sharp vision was a key concept to this lynx-eyed Academia dei Lincei. They were named after the mytho-logical beast, an elusive and ghost-like creature, that sees without being seen. The animal was also known as "the keeper of secrets of the forest," its magic and enigma stemming from the way in which the creature's secrecy was also its strength. The lesson to be learnt from the lynx is that even the lowliest can succeed in life and that the world can unfold itself to those who stop and listen.

Clearly the academy was so named since the lynx was an animal whose sharp vision symbolized the observational prowess needed for science. It was the first academy of sciences to persist in Italy, an early scientific organization that held radical views on science, philosophy, and religion. They studied botany, geology, and zoology. They car-ried out early research in plant and animal dissection, and provided a locus for the incipient scientific revolution, though the Academy was not to become the country's official scientific academy until 1871.

There is also significance in the secret nature of the lynx. The academy had good reason to be elusive in an age of potential religious persecution. Every member of the group was radical. Their number included Giambattista della Porta, a respected elder of the occult, who had written widely on alchemy and magic, and had even been interro-gated by the Inquisition for sorcery. The Flemish doctor Jean Eck was another member of the academy. He had fled his homeland due to his dabbling in black magic and had even been convicted for murder.

The founding president of the Lincean Academy was a young nobleman by the name of Federico Cesi, the second marquis of Monticelli. Cesi was a great admirer of Galileo, since Cesi had an affinity for all those who championed anti-Aristotelian views. But unlike Bruno and those of lowly birth, the Lynxes managed to escape open persecution by the Church, mostly due to Cesi's social position. In 1613 he was made a prince.

After dining with Cesi's academy in April 1611, Galileo was invited to become only the sixth member of this elite society. He proudly accepted. A strong bond was forged between Cesi and Galileo, a bond that was to last until Cesi's untimely death nineteen years later. The academy was to play a dramatic part in the clashes between Galileo and the Church. They provided finance for two of Galileo's most important pending publications, *Letters on Sunspots* (1613) and *The Assayer* (1623). Through his association with the academy, and unlike poor Bruno, Galileo's links to those of noble standing was to provide some limited protection in the troubled times ahead.

Anatomy of the Coming Conflict

Controversy continued to stalk Galileo.

Before we embark in more detail on the growing tension between Galileo and the Church, it is worth delineating an anatomy of the conflict to come. There are three main aspects to the clash, which justify some forethought. Those aspects are personal, political, and religious.

Despite the accolades of wealthy backers, despite the esteem of select and notable fellow academics, Galileo found himself under attack when he returned to Florence late in 1611. Whereas some of the Lynxes secured a degree of immunity by their association with Cesi's high rank, Galileo found himself under increasing pressure. Having friends in high places only served him so far.

There is much danger in a difference of social class. Bruno had held radical, if not extremist, views on science and philosophy. Being from a poor, undistinguished family, he had never achieved established status as an intellectual. Bruno had been burnt at the stake for heresy, his works publicly destroyed, and his bones crushed to powder.

Galileo was playing a dangerous game. He too was of rather modest background. And to make matters worse, like Bruno, he was radical in thought and daring in deed. So too was Cesi. And yet, as a royal figure, Cesi was untouchable.

As far as the personal aspects of the conflict are concerned, much has been written of the character of Galileo himself. Some refer to the "cold, sarcastic presumption,"[25] by which he "managed to spoil his case throughout his life."[26] Others describe him as, "loud, forceful, argumentative and combative,"[27] buoyed up by "a misplaced sense of self-importance."[28]

Scholars in Bologna had set up an anti-Galileo group. The group orbited about the geocentrist astronomer Giovanni Antonio Magini, whose chief beef with Galileo was merely that he had beaten Magini to the chair of mathematics in Padua, a full twenty years before.

In an earlier, published dispute about the use of the proportional compass,[29] Galileo had dubbed an unfortunate competitor "that malevolent enemy of honour and of the whole mankind," "a venom-spitting basilisque," "an educator who bred the young fruit on his poisoned soul with stinking ordure," and "a greedy vulture, swooping at the unborn young to tear its tender limbs to pieces."

You can see the problem. Such intemperate language is the stuff that spawns lifelong enemies and engenders in others a quick contempt. In the quarrel with Ludovico delle Colombe on the discoveries of his telescope, Galileo had considered Colombe's criticisms beneath his dignity. Nonetheless, he could not resist dubbing Colombe "the Pigeon" (*colombe* is dove in Italian) and all his followers "the Pigeon League."

There seems to be little doubt that Galileo's character, along with his radical science and his love of attention, was too incendiary a mixture for many of his enemies, learned and political. He may well have thought himself too smart to end up like Bruno, too infallible in his empirical science, too important in his social connections.

His character apart, Galileo is blameless in wishing to communicate his radical ideas to the public. Galileo loved forceful debate, the more skillful his opponent, the better. In matters of astronomy, Galileo was no doubt aware of the fact that he was right, at least in principle, if not in detail. And his detractors were in the wrong, defending an outdated theory, which was soon to be dispatched to the dustbin of history.

Such opposition to radical ideas was political as well as religious.

When the resurgence of Western Christianity had begun in the tenth century, the universities emerged as centers for the administrative training of the clergy, who had a monopoly on the literate occupations. Those at Paris, 1160, Oxford, 1167, Cambridge, 1209, and Padua, 1222, were typical of these institutions, which acted as repositories of learning, the focus of European intellectual life.

By the later Middle Ages, these universities had become fetters to progress. They had evolved into conservative guardians of the established order, its knowledge, and customs. They were significant barriers to cultural and political advance. For, as Arthur Koestler so forcefully put it:

> There also existed a powerful body of men whose hostility to Galileo never abated: the Aristotelians at the universities. The inertia of the human mind and its resistance to innovation are most clearly demonstrated not, as one might expect, by the ignorant mass—which is easily swayed once its imagination is caught—but by professionals with a vested interest in tradition and in the monopoly of learning. Innovation is a twofold threat to academic mediocrities: it endangers their oracular authority, and it evokes the deeper fear that their whole, laboriously constructed intellectual edifice might collapse. The academic backwoodsmen have been the curse of genius from Aristarchus to Darwin … they stretch, a solid and hostile phalanx of pedantic mediocrities, across the centuries.[30]

It was an opposition of which Galileo was acutely aware. They had browbeaten Copernicus into silence for most of his intellectual life until he published, at the very death. In Galileo's case they made up a resolute rearguard, fixed firmly in preacher's pulpit and professorial chair:

> There remain in opposition to my work some stern defenders of every minute argument of the Peripatetics [the followers of Aristotle]. So far as I can see, their education consisted in being nourished from infancy on the opinion that philosophising is and can be nothing but to make a comprehensive survey of the texts of Aristotle, that from divers passages they may quickly collect and throw together a great number of solutions to any proposed problem. They wish never to raise their eyes from those pages—as if this great book of the universe had been written to be read by nobody but Aristotle, and his eyes had been destined to see for all posterity.[31]

And then, there was the Church itself.

"Out, damn'd spot! Out, I say!"

A new year, and a new controversy. By 1612 Galileo had become caught up in a quarrel with more telling costs. The clash was on the question of sunspots.

It is worth giving a very brief and potted history of sunspots, lest the impression be given that this Renaissance discovery was entirely new. It was not. What *was* new was the interpretation of the phenomenon.

According to *The Book of Han*, completed in 111 AD, ancient Chinese astronomers made reference to spots on the Sun as early as 28 BC. At that time, observers could probably see only the largest spot groups, just as the Sun's glare was diffused by wind-borne dust from central Asian deserts.

A large sunspot group was also seen in 813 AD, at the death of Charlemagne, king of the Frankish Empire that embraced much of western and central Europe at that time. The annals of Charlemagne, as reported by Galileo himself from his history of the Franks, record that eight centuries earlier in France many people had noticed a dark mark on the Sun, lasting eight days. The assembled savants agreed that the spot must have been Mercury. But no transit of Mercury could last even eight hours.

Such blemishes on the face of the Sun were highly significant, not only to superstition, for they would soon become so for science. Much import was read into the visitations of comets at the death of kings, and the same could be said of spots on the surface of the Sun, especially at the passing of an emperor as august as Charlemagne.

But to Galileo, the significance of the spots was the passing of Aristotle, and with him, his cosmology. The very existence of sunspots was a grave problem for Aristotelian celestial physics. Such impurities on the face of the Sun provided great difficulty for a cosmology that held dear the unchanging perfection of the heavens.

Galileo was one of the first Europeans to observe sunspots. As early as the summer of 1609, from the very first days of delving the depths of space with his scope, he had begun to notice the odd blemish on the face of the Sun. In 1611, his solar work started in earnest. Along with his favored disciple, the bright and enthusiastic nobleman Filippo Salviati, Galileo began an intense period of sunspot study.

As the Italians scanned the heavens, a priority dispute was about to brew on the sighting of the spots. Sunspotting had become de rigueur for a small band of European astronomers, equipped with the new telescopic invention. Among this vanguard of early solar scientists were Englishman Thomas Harriot, the Frisian-German Johannes Fabricius, and the rather appropriately named Christoph Scheiner, from Swabia in Germany.

Working at Ingolstadt in Bavaria, Scheiner had taken advantage of a day thick with mist, turned his telescope at the Sun, and discovered "several black drops" on its surface. After a number of such observations, Scheiner had promptly printed his discovery in letter form, under the pseudonym of "Apelles." A copy was sent to Galileo, for his opinion, which is where the problem began.

Now, in a nutshell, the consensus is that Harriot was the first to observe them, in 1610, Fabricius was the first to publish, in 1611, while Galileo and Scheiner also worked on the spots, each man unaware of the others' parallel discovery. Only Galileo claimed priority, the others seemingly uninterested in such a claim.

The trouble was this: Galileo's claim was unsound. Since there was little doubt that Fabricius was the first to publish, Galileo's claim for priority could only be based on his being the first to observe, a year before Harriot had made his own observations. But Galileo could provide no witnesses, no associates, and no correspondents with which to prove his claim. If he had observed at all, he had observed alone.

And, as one commentator put it,[32] Galileo had come to regard telescopic discoveries as his exclusive domain, "it was granted to me alone to discover all the new phenomena in the sky, and nothing to anybody else. This is the truth which neither malice nor envy can suppress."[33]

Nonetheless, Galileo had got his science right. Scheiner, under the guise of Apelles, had begun to popularize the notion that the spots were due to stars moving in front of the Sun. Galileo knew different.

> If I may give my opinion, I shall say that the solar spots are produced and dissolve upon the surface of the Sun and are contiguous to it, while the Sun, rotating upon its axis in about one lunar month, carries them along, perhaps bringing back some of those that are of longer duration than a month, but so changed in shape and pattern that it is not easy for us to recognise them.[34]

In January 1612, Christoph Scheiner published his book *Three Letters on Solar Spots*, once more under the pseudonym of Apelles. Now Scheiner was a Jesuit astronomer of great repute, and any man who chose to launch into a dispute with him did so at the risk of turning the order against him.

Galileo was well aware of the tricky times he himself was about to face. Before the publication of his own book, *History and Demonstrations about Sunspots and their Properties*, in January 1613, Galileo wrote to Cesi at the Lincean Academy, predicting trouble: "I expect the tempest over the mountains of the Moon will be a joke compared to the lashings I will receive over these clouds."[35]

He was not to be disappointed. The reception was bound to be hostile. For, not only did Galileo's book show convincingly that Scheiner was wrong but it also proved Aristotle to be in error. The Sun, like the Moon, was subject to degradation and decay.

The book held a further feature to inspire fury in Aristotle's followers. It contained Galileo's first printed support of the Copernican system. Up to the age of fifty, Galileo had spent most of his life, either keeping his opinion to himself, or else confining his pro-Copernican remarks to polite after-dinner conversations. They were never in print, until now.

The actual passage where he mentions Copernicus and his system is rather innocuous and begins with a further reference to Galileo's mistaken notion that Saturn has moons: "And perhaps this planet also, no less than horned Venus, harmonises admirably with the great Copernican system, to the universal revelation of which doctrine propitious breezes are now seen to be directed towards us, leaving little fear of clouds or crosswinds."[36]

Galileo's sunspot script drew cant from all quarters. The clergy were appalled. They were horror-struck by the fact that Galileo and the Lincean Academy had chosen to publish this book of radical science in the vernacular, enabling ordinary people to read its impious pages. The Aristotelians were greatly aggravated. They had been completely outmanoeuvred by Galileo, who had deftly destroyed Aristotle, while keeping clear of contradicting Scripture. After all, there was no mention of sunspots in the Bible.

THE CREEPING CENSURE

Dark clouds of oppression were looming. The first serious attack on Copericanism for religious reasons came from a high-flying Aristotelian, the leader of the Pigeon League. In his treatise, *Against the Motion of the Earth*, Ludovico delle Colombe quoted Holy Scripture to show that the Earth was fixed at the center of God's domain.

And when Galileo's star student, Benedetto Castelli, was appointed as a mathematician to the University of Pisa, replacing Galileo, he was explicitly banned by the chancellor of the university from teaching the motion of the Earth. The Chancellor, Arturo d'Elci, was another member of the Pigeon League and an obsessive follower of Aristotle.

Aware of this creeping censure, Galileo had cautiously sought opinion on Holy Scripture before publishing his discourse on sunspots.

He had written to Cardinal Conti, Prefect of the Roman Inquisition, admitting that his theory was anti-Aristotelian, but enquiring as to whether or not it was antidoctrinal.

The Cardinal responded in favor of Galileo. On the question of the immutability of the skies, Scripture found for Galileo, rather than Aristotle, said Conti. As for Copernicanism in general, that was a different matter. Conti's opinion was that the annual motion of the Earth proved no problem, but the daily rotation jarred with some passages, unless interpretation was liberal. And such liberal readings of Scripture were only likely to be tolerated should convincing evidence of the Earth's motion be made readily available.

Tricky.

Conti's judgement was only a temporary respite, especially since Galileo had no readily available evidence of the Earth's motion. Elsewhere, the attacks continued. While staying in Florence with his friend Filippo Salviati (whom Galileo made famous in his two great *Dialogues*), news came that a Dominican monk, Niccolò Lorini, another member of the infamous Pigeon League, had used references to Scripture in fiercely critical sermons against Galileo.

Galileo wrote to Lorini, demanding an explanation. He received a rather devious and disappointing response.

I have never dreamt of getting involved in such matters … I am at a loss to know what grounds there can be for such a suspicion, as this thing has never occurred to me. It is indeed true that I, not with a desire to argue, but merely to avoid giving the impression of a blockhead when the discussion was started by others, did say a few words just to show I was alive. I said, as I still say, that this opinion of Ipernicus—or whatever his name is—would appear to be hostile to divine Scripture. But it is of little consequence to me, for I have other things to do.[37]

Ipernicus, indeed.

But Niccolò Lorini wasn't as innocent as he made out. By February of 1615, Lorini had filed a written complaint with the Inquisition against Galileo's Copernican views. The main item of evidence in support of his case was a letter defending the Copernican theory that Galileo had written to his friend Benedetto Castelli, who recently had a run-in with one Grand Duchess Dowager Christina of Lorraine.

This protest to the Inquisition had come hot on the heels of other minor public scandals. Preaching a sermon, "Ye Men of Galilee, why stand ye gazing up into the heaven?" in the Florentian church of Santa Maria Novella, the Dominican monk Father Thommaso Caccini attacked Galileo and other mathematicians who supported the Copernican view as heretics.

The Bishop of Fiesole had gone further. He demanded that Copernicus himself be instantly jailed, but was stunned to learn that he had been dead for seventy years. At first, in the face of such growing hysteria, a kind of truce was called. Caccini's superior apologized to Galileo in writing, and Lorini's complaint to the Inquisition dropped, at least for the time being.

But the damage was done. Galileo's letters to Castelli and the Grand Duchess Christina were dynamite. True, they were so carefully worded they could not be condemned as heresy. However, their intent was beyond doubt. They constituted a challenge to the authority of the Holy Scripture, and of the Holy Church itself. They remained on the files of the Inquisition and in the mind of one man in particular.

THE "HAMMER OF THE HERETICS"

Cardinal Robert Bellarmine. If there were one man who could be described as Galileo's main adversary during this historic controversy, it would be this "Hammer of the Heretics." He was one of the men responsible for burning Bruno at the stake for his obstinate heresy only a dozen or so years before. In England, he was thought to have been the mastermind behind the Gunpowder Plot in 1605, an act for which he was described as being a "furious and devilish Jesuit." He was canonized for his work in 1930.

Bellarmine was a worthy opponent.

A man of considerable culture, he had been professor of theology until 1589, when his talents as a future theologian, prepared to defy popes and kings, was first realized. Neither was he ignorant of astronomy. Having accepted a commission from Pope Gregory XIII to lecture on polemical theology in the new Roman College, one of Bellarmine's first duties was as "Master of Controversial Questions." There, he was close to the cluster of Jesuit astronomers in Rome who had been among the first won over to Copernicanism by Galileo's exciting sightings with the spyglass.

So Cardinal Robert Bellarmine was no *ignorante*. He was a fanatic. He was motivated by a supreme vision: that of the Universal Church, a theological super-state. It was an idea out of time. For political, economic, and scientific developments throughout Western Europe militated against such a super-state. Not only was there the Protestant heresy to deal with, but also the rising tide of nationalist tendencies derived from burgeoning trade and the promise of the new science.

Bellarmine's legacy is as one of the great controversialists of history. After all, at this time when the power of the papacy was under pressure, Bellarmine had plenty of practice in polemics. He became involved in the Galileo affair when, in 1615, he wrote to both Galileo and Father Paolo Antonio Foscarini, a man who had supported Galileo publicly by publishing a book that took the dangerous step of trying to reconcile Copernicus with Scripture.

The letter from Bellarmine began by suggesting that heliocentric ideas were "a very dangerous thing, not only by irritating all the philosophers and scholastic theologians [i.e., the Aristotelians], but also by injuring our Holy Faith and rendering the Holy Scriptures false." However, Bellarmine admitted that if there were positive proof, "then it would be necessary to proceed with great caution in explaining the passages of Scripture which seemed contrary, and we would rather have to say that we did not understand them than to say that something was false which has been demonstrated."[38]

Galileo was in a difficult position. Bellarmine would tolerate heliocentrism, as long as it was treated purely as hypothesis, and not as absolute truth unless there was conclusive proof. Galileo *had* no "conclusive" proof. Sure, he had plenty of powerful arguments and had amassed many suggestive observations to support his position. But, as with all radically new scientific work, Copernicus's creation was an unfinished symphony, a score with copious errors in timing and detail.

Prepared for battle, Galileo traveled to Rome in December of 1615. His mission: to convince Bellarmine, the Inquisition, and ultimately the papacy, of the veracity of the Copernican case. His decision to travel was not a popular one. The Tuscan ambassador in Rome had already warned Galileo's boss, the Grand Duke of Tuscany. Now he reported:

> He is passionately involved in this quarrel, as if it were his own business, and he does not see and sense what it would comport; so that he will be snared in it, and will get himself into danger, together with anyone who seconds him ... For he is vehement and is all fixed and impassioned in this affair, so that it is impossible, if you have him around, to escape from his hands. And this is a business which is not a joke, but may become of great consequence, and this man is here under our protection and responsibility.[39]

If Galileo's arrival in Rome spelt trouble, the "proof" of the Earth's motion he brought with him spelled worse. This proof was a theory of the tides. Rather than the theory providing decisive evidence of the Copernican case, it was a dead end, which was to prove only this: that Bellarmine and the papacy had reached their limit on open discussion and freedom of thought.

Now Kepler had already produced a correct theory of the tides. It was somewhat hidden away in his *Astronomia Nova*, published in 1609, and contained the results of Kepler's decade-long calculations on the motion of Mars. At over 650 pages in English translation, this massive tome is a difficult enough work from a modern perspective, laced throughout as it is with the mumbo-jumbo of the mercurial mystic.

We spoke earlier in this chapter of the cryptic correspondence between Kepler and Galileo, and also intimated that Galileo's ignorance of Kepler's work was to cost him dearly with the Church. So it was with the theory of tides. It seems that, for someone of Galileo's legendary impatience, the book was too long, its context too obscure.

Galileo replaced Kepler's explanation of a correct theory based on the Moon's attraction with a declaration that the tides were a direct result of the Earth's combined motions, first, about the Sun, and second, about its own axis. According to Galileo, these combined motions meant that the oceans moved at a different speed from the land.

This theory was quite out of character for Galileo. For one thing, his theory of the tides claimed only one high tide per day, on the stroke of noon. But such a practical empiricist as Galileo surely would have known that there were daily two high tides and that they crept around the clock, day by day. It must surely be Galileo's greatest blunder, analogous to the famous "mistake" Einstein made in figuring a "cosmological constant" into his field equation for General Relativity.[40]

The theory was an act of desperation. Bellarmine must have been well aware of the weakness of Galileo's case. He held a brief discussion with the pope (Paul V Borghese), a man who abhorred "the liberal arts and his [Galileo's] kind of mind."[41] They decided that Galileo's stance was "erroneous and heretical."[42]

So the decree came. Bellarmine summoned Galileo to his residence and warned him not to hold or defend the Copernican theory. On February 23, 1616, the Qualifiers of the Holy Office declared that heliocentrism was "absurd in philosophy and formally heretical,"[43] and that the notion that the Earth has an annual motion was "absurd in philosophy and at least erroneous in theology."[44]

On March 5, a more moderate version of the Qualifiers' position was eventually published. Copernicus's *De Revolutionibus* was placed on the Index of Forbidden Books, subject to revision. A further unsigned transcript in the Inquisition files, not discovered until 1633, also states that Galileo was particularly forbidden "to hold, teach, or defend in any way whatsoever, verbally or in writing" the Copernican doctrine.[45] For the next few years, Galileo had to watch his step.

The "Dialogue"

After the decree, Galileo returned to Florence.

For seven years he published nothing. Meanwhile, in 1618, three different comets appeared in the skies of Western Europe. Many, including Archduke Leopold of Austria, sought Galileo's views on these elusive visitors of the sky. So Galileo, cautiously, drafted a series of lectures, which mistakenly held to the Aristotelian notion of comets moving on straight lines through the Earth's upper atmosphere. The lecture was delivered by one of his students.

In 1624, Galileo had six long audiences with Pope Urban VIII over the course of a six-week period. The subject of those audiences has long been a matter of some debate. Perhaps the only point that has been established with any conviction is that the pope assured Galileo that he could write about the Copernican theory, as long as he treated it as a mathematical hypothesis. Supposing he now had papal blessing, and perhaps believing too much in the healing nature of time, Galileo, now past his sixtieth year, began his great defence of Copernicus.

By January 1630 it was complete. Galileo's *Dialogue Concerning the Two Chief World Systems* is a discourse between three characters, carried out over four days, and written in Italian. From the outset of the *Dialogue*, the very naming of its three main characters speaks volumes of the stance that Galileo was to adopt in the rest of the book. Salviati, the brilliant academician, is Galileo's advocate. Sagredo, the intelligent layman, is, at least initially, neutral. And Simplicio, dedicated follower of Aristotle, Ptolemy, and therefore the Church, is the *ignorante*, the well-meaning muggins, who understands little of the world in which he lives. It is hardly surprising that the Inquisition would object to this last name's resemblance to "simpleton."

Salviati and Sagredo, were named after much loved friends of Galileo, both now dead. Simplicio, Galileo claimed, derived his name from Simplicius, the sixth-century Neoplatonist philosopher. But Simplicio is far more likely to have been based both on delle Colombe, he of the Pigeon League, and Cesare Cremonini, the Paduan colleague of Galileo's, who refused to look through the spyglass.

Sadly, the book is greatly flawed.

Galileo again exhibits complete ignorance of the planetary theories of Kepler, written a full score years before. It is almost as if he neglected his research in astronomy, directly after his initial use of the telescope, to focus on the conflict at hand, the propaganda crusade against the Aristotelians. Witness the facts that, for instance, Galileo seems totally unaware of Kepler's laws of planetary motion, had no idea that Kepler had solved the problem of Mars's wayward orbit, and had no notion of Kepler's mathematical discovery that the planets moved about the central Sun in ellipses rather than perfect circles.

Thankfully, the *Dialogue* is not limited to the thoughts sprung by Galileo's use of the telescope. It ranges through much of contemporary science, from a discussion of magnetism, to early thought experiments on the question of relativity. But even most of these often beautifully written passages form part of the main thrust of the book: a refutation of the Aristotelian view of the cosmos.

And that's how the book really makes its mark. The eponymous chief world systems are that of Copernicus and Ptolemy. The *Dialogue* does not deal with Tycho's system, which had fast become the preferred system of leading scholars who were too gutless to embrace Copernicus. Tycho's system, Galileo believed, was a sorry compromise. So, best aim at the main prize of Ptolemy, for with him you get Aristotle, and if you hit your target, the whole system comes crashing down, Church and all.

The Trial of Galileo

Finally, the axe fell.

The first printed copies of the *Dialogue* came off the press early in 1632. Though Italy was torn apart by plague, news traveled fast enough to Rome, where Pope Urban and the Holy Office had the book quickly seized. By October, Galileo was summoned, on the basis that the *Dialogue* contradicted in spirit and letter the decree of 1616.

Galileo stalled. For four months he kept them waiting in Rome, while he delayed his journey on grounds of sickness and various other ploys. But he was merely putting off the inevitable. The first interrogation at the Holy Office took place on April 12, 1633. A typical inquisitorial tactic was employed: Galileo was asked whether he could guess on what grounds he was being summoned.

He guessed correctly, it was the *Dialogue*. Then, rather unbelievably, when he was questioned as to whether he had disobeyed the decree neither to "hold, defend, nor teach" the case that the Earth moves and the Sun is stationary, Galileo's bizarre response was to affirm the contrary, that his book "demonstrated the opposite of the Copernican opinion"[46] and that it showed the case of Copernicus was "weak and not conclusive."[47] And with that, the first hearing ended.

But secretly, behind closed doors, the die was already cast. Only a matter of days after the hearing, three moderators of the Inquisition, who had been selected to scour the *Dialogue*'s contents, submitted their reports and their verdict: Galileo was guilty. Guilty not just of discussing the Copernican case as hypothetical. But also guilty of teaching, defending, and holding Copernicanism as true.

It is not as if Galileo was particularly subtle about his case. The *Dialogue* does not mince its words in describing those who do not share the Copernican view. Such people are referred to variously as "hardly deserving to be called human beings," "dumb idiots," and "mental pygmies." Indeed, the moderators themselves may have felt similarly inflicted in their reading of the book, should they not share Galileo's opinion.

So Galileo was called to private interview. On April 28, 1633, he met with Father Vinco da Firenzuola, the commissary of the Inquisition. Firenzuola's intention was to save appearances, but mostly to uphold the reputation of the Inquisition. He proposed that Galileo recognize and admit he had gone too far in his book, and "that done, he might have his house assigned to him as a prison,"[48] a solution suggested by one of the trial judges, Cardinal Francesco Barberini, the pope's brother.

Two days later, Galileo was called again. The second formal hearing was a changed encounter. Galileo made a formal statement, in which he claimed, "I have not held and do not hold as true the opinion which has been condemned, of the motion of the Earth and stability of the Sun."[49] He was clearly terrified. He was seventy years of age, and despite his numerous friends among the merchants and the clergy, he was exposed and alone. Without the benefit of Bruno's courage, or Kepler's researches, his attempt at single-handedly outwitting the Church had come to this. He was a broken man.

Feeling vulnerable, Galileo even made an offer to add another couple of "days" to the *Dialogue* so he could "correct" the arguments therein. Ten days later he was called again. On May 10, he submitted his written defence, begging for leniency from "the most Eminent Lords, my judges"[50] and claiming that "those faults which are seen scattered through my book ... have inadvertently fallen from my pen."[51]

The record of the trial serves to emphasize in fascinating contrasts of light and dark, illumination and shade, the contradictory attitude of the Holy Church to Galileo. Ultimately, they condemned him as a heretic, wishing to make of Galileo an example of all those who use the new empirical science, and who dare to question the authority of the Church.

But they also deeply feared Galileo and the new astronomy. Throughout the trial's proceedings Galileo was treated with great courtesy: confined to the opulent quarters of the Holy Office itself, rather than languishing in the Inquisition's dungeons; dining on the Church's food and wine, rather than the prison fare Bruno had to contend with.

So the condemnation of Galileo was a delicate business. The influence of this world-famous scholar, with his many powerful friends, had to be balanced against the wider concerns of the power of the Church. It was not merely Galileo's refusal to obey the 1616 decree of the Holy Office that worried the Church. The new astronomy held many concerns for an office wishing to maintain its longstanding religious and political influence in Europe. As Thomas Kuhn eloquently put it:

> When it was taken seriously, Copernicus' proposal raised many gigantic problems for the believing Christian. If, for example, the Earth were merely one of six planets, how were the stories of the Fall and of the Salvation, with their immense bearing on Christian life, to be preserved? If there were other bodies essentially like the Earth, God's goodness would surely necessitate that they, too, be inhabited. But if there were men on other planets, how could they be descendents of Adam and Eve, and how could they have inherited the original sin, which explains man's otherwise incomprehensible travail on an Earth made for him by a good and omnipotent deity? Again, how could men on other planets know of the Saviour who opened to them the possibility of eternal life? Or, if the Earth is a planet and therefore a celestial body located away from the centre of the universe, what becomes of man's intermediate but focal position between the devils and the angels? If the Earth, as a planet, participates in the nature of celestial bodies, it can not be a sink of iniquity from which man will long to escape to the divine purity of the heavens. Nor can the heavens be a suitable abode for God if they participate in the evils and imperfection so clearly visible on a planetary Earth. Worst of all, if the universe is infinite, as many later Copernicans thought, where can God's Throne be located? In an infinite universe, how is man to find God or God man?[52]

Galileo was called for his final examination in June of 1633. After taking oath, he was asked for his true conviction concerning the two chief world systems. His answer was, "Assured of the wisdom of the authorities, I ceased to have any doubt; and I held, as I still hold, as most true and indisputable the opinion of Ptolemy, that is to say, the stability of the Earth."[53] And on repeated questioning, he repeated the same refrain, "I do not hold, and have not held, this opinion of Copernicus since the command was intimated to me that I must abandoned it."[54] Following Galileo's last answer, the trail minutes read, "And as nothing further could be done in execution of the decree, his signature was obtained to his deposition and he was sent back."[55] And so the imprisonment began.

Once more the curious and contradictory nature of the trial shines down the years. To some extent Galileo's behavior was condoned. Both the Inquisition and the astronomer knew that he was lying. Lying in his contention that he had written the book against the heliocentric cause, and lying in that he had presented the case of Copernicus as more than merely probable.

But the Holy Office still had the new astronomy in its sights. The final verdict on the work of Galileo was not as lenient or condoning. In an action that was not announced at the time, the publication of anything else he had written or ever might write was also banned. Not even a scholar of the status of Galileo was to mock the Church. Eventually, he returned to Florence, under house arrest for the rest of his days.

The *Dialogue* was smuggled out of Rome shortly after the trial. Translated into Latin, it was printed in 1635 and spread like wildfire throughout Europe. The following year, the manuscript of Galileo's other great contribution to the revolution was also smuggled out into the light. His treatise on dynamics, *Discourses and Mathematical Demonstrations Relating to Two New Sciences*, was Galileo's final book and a testament to his work in physics of the preceding thirty years.

In the course of his closing years he received a succession of eminent guests to his home in Florence. Sadly, his eyesight finally fading, his working life was over. As he wrote to his faithful friend Elia Diodati,

Alas, your friend and servant Galileo has been for the last month hopelessly blind; so that this heaven, this Earth, this universe, which I, by marvellous discoveries and clear demonstrations, have enlarged a hundred thousand times beyond the belief of the wise men of bygone ages, henceforward for me is shrunk into such small space as is filled by my own bodily sensations.[56]

CHAPTER 9

THE "DARWIN" AFTERMATH

2009 was Darwin's year. Throughout the world, but above all in Britain, national programs of celebration were organized to commemorate Charles Darwin's life, times, scientific ideas, and their impact. The Darwin200 venerations were especially heady between his 200th birthday, on February 12, and the 150th anniversary of the publication of the *Origin of Species*, on November 24.

Many organizations and agencies, large and small, were involved. They ranged from arts councils and the BBC, through Westminster and the Woodland Trust, to various universities and zoological societies, up and down the land. They planned and executed an incredible array of events to honor the life and legacy of a single biologist: conferences and exhibitions, graphic novels and dramatic performances, and documentaries on radio and television.

The doctrine of Darwinism was on display. In just one week in February 2009, it was used to weave an evolutionary analysis of literature, to navigate links from Darwin to online social networks, to fool around with the idea of God as an evolutionary adaptation, and to analyze Darwin's impact on art and aesthetic theory. Perhaps most tellingly, there was also a major feature looking at the "need" to revisit early efforts at Darwinian economics.

Darwin is big business.

With over three and a half million visitors a year, the Natural History Museum in London was a keystone in the Darwin celebrations. Originating from collections within the British Museum, and designed in the landmark Romanesque style of English Gothic architect Alfred Waterhouse, the Natural History Museum was opened in 1881, at the peak of the Victorian obsession with "Darwinia." One such curiosity

is a two-and-a-half tonne white marble statue of Darwin, created by
Sir Joseph Boehm and unveiled to the public in June of 1885. In
1927 it was moved to make way for an Indian elephant specimen and
then moved again in 1970 to the North Hall. But as part of the early
preparations for the coming anniversary, in May 2008 the statue was
returned to its original prime position, at the top of the main staircase
in the museum's iconic Central Hall. Darwin looks down upon his
domain once more.

The museum's showcase for Darwin200 was the Darwin Big
Idea Exhibition. Billed as the biggest ever exhibition about Charles
Darwin, it celebrated "the impact of the revolutionary theory that
changed our understanding of the world." Revealingly, the exhibition's campaign poster appeals to the faithful with the absurd words,
"If you had an idea that was going to outrage society, would you keep
it to yourself?"

But the truth is far more intricate. The theory of evolution is not
the work of a single "genius."

Since antiquity, some of the finest philosophers have wondered
at the rich variety of life. During that time, divine creation had not
always been thought of as the causal factor. A stream of very able
thinkers, running from Empedocles, Epicurus, and Lucretius through
to Leonardo da Vinci, had tended toward a more secular speculation.
Instead of believing in the Great Chain of Being, that every form of
life was created by God and remained unchanged since, these sages
also looked to nature's inherent patterning for a cause.

So it was with luminaries such as Francis Bacon, René Descartes,
and Gottfried Wilhelm Leibniz, all of whom favoured natural over
supernatural causes for species change. The French naturalist Étienne
Geoffroy Saint-Hilaire had made significant progress in the late
eighteenth century. Saint-Hilaire had suggested that some species adapt
and survive when environments change. Others perished and became
extinct. Saint-Hilaire had come the closest so far to understanding the
process of evolutionary change. But he lacked that all-important explanation, of how new species formed, and how the millions of species on
planet Earth had emerged from a common ancestor.

Suppose that poster from the Natural History Museum is spot on.
Suppose that evolution "was the biggest revolution in scientific thought
the world would ever see," not that they could ever know, of course.
But the key to the revolution was in identifying the "missing link" in the
theory of evolution, to recognize natural selection as the mechanism of
change. Many brilliant brains had been tried and found wanting. Even
Erasmus Darwin had failed to spot this particular missing link.

So how was it that a rather conventional and unimaginative naturalist from the quiet town of Shrewsbury in England led this revolution? True, he had spent his early years rather rebelliously. But he soon "bore the weight of expectation willingly,"[1] and after a single, very well-publicised and long voyage, he traveled rarely, crippled in Kent with illness and panic disorders of social and agoraphobias. Darwin certainly did not inherit his grandfather's boldness of spirit. And as the English philosopher A. N. Whitehead said, "Darwin is truly great, but he is the dullest great man I can think of."[2]

But Britain could brag of another brilliant biologist. Younger than Darwin by a decade, this naturalist was radical and open-minded, an autodidact who was both intuitive and unconventional. The younger man roamed freely about the globe, from the rainforests of the Amazon Basin, to the Malay Archipelago. Here he identified the Line that divides Indonesia into two distinct parts. One in which creatures closely related to those of Australia are common, and one in which the species are almost exclusively Asian in origin.

By the 1850s, Alfred Russel Wallace, like Darwin, had also sailed the world. He too had carried out an exhaustive program of observation and research. He too had settled into considering his collected data. And he too was recognized by his peers as one of the foremost workers in his field. But the social background Wallace brought to bear on his studies differed dramatically.

Darwin's life was one of comparative luxury. He belonged to a social class that was wealthy and privileged, a family that was very well educated and associated with some of the key players of Victorian science and society. Wallace, on the other hand, worked for a living. He had little to speak of in terms of social influence and connections, had left school at the age of fourteen, and suffered constant financial struggle throughout his long and productive life.

The conventional account of the Darwin-Wallace story has a reluctant Darwin, badgered by Lyell, laboring over an exhaustive account of the origin of species through natural selection. Darwin had intended the book to be lengthy and to take years of toil. Perhaps, like Copernicus, he secretly wished he would not have to live to read the reviews, religious and scientific.

Suddenly, on June 3, 1858, so the story goes, when Darwin had penned practically only his preamble, everything changed. A letter postmarked the Malay Archipelago arrived at Darwin's door. Wallace's communication contained an early draft of an essay titled "On the Tendency of Varieties to Depart Indefinitely from the Original Type." And Wallace wanted Darwin's feedback on the paper.

The usual tale continues with Darwin reacting with stunned shock, since the theory in Wallace's letter was identical to his own. As Darwin wrote to Lyell that afternoon, "I never saw a more striking coincidence."[3] Wallace had been recovering from malaria when "it suddenly flashed upon me that ... in every generation the inferior would inevitably be killed off and the superior would remain—that is, the fittest would survive."[4] Wallace had drafted the theory in a whirlwind of activity. In only three days he had it penned and sent to Darwin, who was well-known in scientific circles for backing the theory of evolution.

The conventional account concludes with a compromise. Darwin at first wanted to do the decent thing, let go his priority, and give all the glory to Wallace. "I should rather burn my whole book, than that he or any other many should think that I had behaved in a paltry spirit,"[5] he told Lyell. But Lyell and Hooker differed. They induced Darwin to publish a joint paper, outlining his and Wallace's conclusions. Within a year came Darwin's book.

The Origin of Species is rather bloodless. Copernicus had subjected his readers to bouts of ecstasy on the subject of the Sun. Galileo had gone into fiery rapture in his dialogues of dynamics and world systems. And Newton had weaved the works of God into his great philosophy of universal gravitation. Not so Darwin. The sum total of the *Origin*'s two hundred thousand words is a reading relentless in facts and figures.

DISSING DARWIN

But there is the devil in the detail. There is a more skeptical account of the same story of how Darwin's *Origin* came to be published. On the centenary celebration of *Origin* in 1959, Cyril Dean Darlington, Sherardian Professor of Botany at Oxford University, began to wonder why there seemed to be no apparent original germ of the idea of natural selection in Darwin's work.

Despite Darlington's clear enthusiasm for "Darwinism," he was puzzled that the *Origin* itself contained no account, within its abundant pages, of how Darwin had come by his theory. Like never before, Darlington went against the stream of conventional academic current, confronting the matter rather candidly:

> How is it, we may now ask ourselves, that so much obscurity overhangs the development of the greatest of modern ideas? After a hundred years we are almost as uncertain of the authorship or editorship of Darwin's writings as we are of those attributed to Homer or Hippocrates.

This is due, on the one hand, to the fact that people who investigate the history of science are historians who are not entirely clear about the meaning of its ideas. They also often believe what the discoverer writes about his own discoveries, which, as we see, is not a wise thing to do. On the other hand, among scientists there is a natural feeling that one of the greatest of our figures should not be dissected, at least by one of us. The myth should be respected.[6]

More scholars became skeptical. Writing at the same time as Darlington, Loren Eisley, professor of Anthropology and the History of Science at the University of Pennsylvania, was not convinced by Darwin's account of how he came to find the key to his theory. Darwin had suggested that Thomas Malthus had given him the idea of natural selection. But Eisley began to suspect the Malthus alibi would not wash.

A science writer of some repute and a lifelong Darwin supporter, Eisley dug deeper. By studying archival records of *The Annals and Magazine of Natural History*, Darwin's favorite scientific journal, Eisley became convinced of a counter-theory. Namely, that Darwin had taken the idea of natural selection from English zoologist Edward Blyth, who had written on the topic between 1835 and 1837.

Eisley's logic is captivating, especially since it gives an insight into the Christian and creationist tendencies of Darwin's early thoughts on evolution.

The idea of deep time, or geologic time, had first been raised in the eleventh century. Both Persian geologist and polymath Avicenna (Ibn Sina) and Chinese naturalist and polymath Shen Kuo had identified the Earth with an exceedingly long history of change and development. In the west the concept had been progressed in the 1700s by James Hutton, though Darwin, of course, had come across it by reading Lyell.

But despite Lyell's insistence on "deep time," Darwin still believed in an Earth history of thousands, rather than millions, of years. He was sold on the notion of quick organic change. If species did change from one form into another, Darwin supposed, there would be no time for gradual change. Rather, Darwin dealt up change in biblical and catastrophic quantities. An organism would leap from one form into another, *per saltum* (at one bound), to better fit the new surroundings. In this picture of evolution, an organism that was slow to adapt would expire in the alien environment. And a species could survive only if it changed in an instant into a different form.

In Eisley's view, Darwin's early theories bear little resemblance to the processes he described eighteen years later in the *Origin*. Eisley was also acutely aware of the blind eye most academics turned toward this important conversion.

> The fact is that an important shift in Darwin's thinking remains undocumented. It has not even been discussed. In the *Origin,* slow and imperceptible transformations extending over vast ages of time have replaced his early and immature speculations on organic change *per saltum.*[7]

What had brought about the change in Darwin? Eisley believed it was Blyth. By rooting through Darwin's trial essays in the decade or so before the *Origin,* Eisley was convinced that Darwin had taken the idea of natural selection from Blyth, without attribution.[8] The examples given above, of the skeptical work of Darlington and Eisley, stem from the centenary celebrations of the *Origin.* They are mere surface compared to the depth of detail that was published on the eve of Darwin200 by Roy Davies, a BBC producer specialising in landmark investigative historical documentaries that challenge popular beliefs.

THE WALLACE DYNAMIC

The central theme of Davies' book on Darwin is Alfred Russel Wallace.[9] Davies' contention is that Darwin deceived his way into becoming the foremost evolutionary theorist at the time and has since been mistakenly identified throughout the world as arguably one of the most influential theorists in the history of science and civilization. And the main victim of this perpetration by Darwin was Wallace.

According to Davies' account, Darwin's accomplices in this deception were his friends and colleagues Joseph Hooker and Charles Lyell. Both men "agreed to put their own reputations on the line."[10] True, they read a joint Darwin-Wallace paper at a meeting on July 1, 1858, for the Linnean Society of London, the world's premier society for the communication of natural history. But in reality they were party to an unseemly collaboration "to ensure priority for Darwin,"[11] their enduring friend of rank and social standing. They manipulated key dates and events, making it look like Darwin had indeed devised the key features of adaptation and natural selection, independently of Wallace, who at the time was collecting specimens in the Malay Archipelago.

Wallace was told of the Linnean reading. But the account he heard was economical with the details. Uninformed of most of the machinations, Wallace was flattered to feature at all in such a rarefied atmosphere as that of the Linnean elite. Being of more middling means and lower social class, Wallace would never have made it alone. But then, this whole affair was not about Wallace. It was about Darwin.

In truth, the Linnean reading was forced onto the Society agenda in great haste. For Darwin had received the infamous Ternate letter from Wallace, setting out the mechanical details of natural selection. And the whole scheme was a plot to prevent almost certain priority for Wallace, unless something was done for Darwin.

It is a fascinating case study in the history of science communication. It depends on what Wallace communicated to Darwin and when. In Davies' words, "They [Lyell and Hooker] agreed, as Darwin was now claiming, that he had sketched out his evolutionary theory not in 1842, but in 1839. Moreover, they claimed that the contents of the 1844 essay [Darwin's first draft of a theory] had not only been read by Hooker, but had been communicated to Lyell himself. There was no mention of the fact that both men had voiced serious objections over several years to Darwin's migration theory, which (alongside the idea of perfect adaptation) had been central to his thinking in 1844 and for a long time after."[12]

Davies conjures up a painstaking analysis of the Victorian postal and shipping records, and through his examination tries to demonstrate that Darwin was creative about the true dates of communications from Wallace. By either claiming that critical letters were delayed, or by simply destroying some, Darwin aimed, says Davies, to establish himself as the primary evolutionary theorist. But in truth, Wallace provided the bullets that fired Darwinian evolution.

What drove Darwin to spin such a web of duplicity? Wallace's crucial Ternate paper was the decider. His *On the Tendency of Varieties to Depart Indefinitely from the Original Type*, dated "Ternate, 1858," is the crux of the whole affair. It had a startling effect on Darwin. But not the kind outlined in the conventional account. Instead, Davies portrays "a very secretive man with a driving ambition."[13] In short, "Had Alfred Russel Wallace sent his letter of March 1858 not to Charles Darwin but to the editor of the *Annals and Magazine of Natural History*," says Davies, "it is likely that we would today talk about Wallaceism rather than Darwinism."[14]

Such is the intrigue thrown up by the 150th anniversary of the *Origin of Species*. After Darwin's death, a small industry sprang up. With biographies penned by the patriarch's family, the Darwin myth was created, one that was jealously guarded and carefully crafted. Academic scholars followed suit. Few wanted to hear any evidence against their reputed hero.

But Darwin wrote around fourteen thousand letters. And, Davies says, Darwin's meticulous filing system means that more skeptical scholars can reassess the case. Fifty years ago, anomalies and coincidences began to be uncovered, allowing them to question Darwin's probity and ethical behavior in relation to Wallace, in particular.

Davies comes to a stunning conclusion. From the details developed by such skeptics over the last half-century, and the new evidence he unearthed from Victorian postal and shipping records, Davies says there is a compelling case to be made against Darwin. Any reasonable person would "conclude it is likely he [Darwin] committed one of the greatest thefts of intellectual property in the history of science."[15]

AN ANATOMY OF SELECTION

Suppose skeptics such as Darlington, Eisley, and Davies are correct. There certainly seems to be a strong case to be considered, especially given the detail of the duplicity demonstrated in Davies' book. Perhaps it should not be surprising that scholars and academics have failed to examine the case with anything like sufficient thoroughness. After all, there are many, perhaps the vast majority, who consider Darwin to be one of Britain's greatest scientific figures and national heroes.

How is the skeptical case to be explained? And how did the Darwin-Wallace affair crystallize within the anatomy of Victorian society, both before and after the publication of Darwin's *Origin of Species*?

Davies himself cites several deciding factors.

One is national pride. Davies points out that most studies, bar his, that questions the authenticity of Darwin's theory was American, not British. So it was with Eisley. And so it was, too, with other skeptical accounts. Distinguished U.S. psychologist Howard Gruber discovered that Darwin completely rewrote his Galapagos entries to take in the new ideas he had gleaned from the islands' vice governor. Others, such as Yale graduate student Lewis McKinney and Yale professor Leonard Wilson, presented research that strongly portrayed Wallace as the leading player in the unfolding drama of evolutionary theory.

Another factor is social class. A number of scholars have suggested there are two main men at the center of this story of evolution. So why praise Darwin to the detriment of Wallace? The answer has much to do with the way in which the science establishment is embedded within the British class structure. Davies' take on the matter is as follows:

> Although the Linnean Society meeting judged Darwin and Wallace to be equally worthy of recognition as the originators of the theory

of evolution, one of them had to be recognised as pre-eminent. The Linnean Society was made up of gentlemen natural philosophers. Wallace, who had written out the complete theory of evolution to which they had listened in silence, was not a gentleman. Charles Darwin, whose unconnected thoughts were contained in two extracts from a 14-year-old essay and a copy of a recent letter, was a gentleman. In the social context of the time, a gentleman always trumps an employee. Thus, the document merging the two presentations referred to the "Darwin-Wallace" theory of evolution. In the lifetimes of both men (Darwin died in 1881, Wallace in 1913) it was usual for the theory to be referred to by this title, but after Wallace's death it became "Darwin's theory of evolution". For almost a century, Wallace's scientific achievement has been effectively buried under Darwin's reputation.[16]

Indeed, though of course one could set this picture in a broader frame. From the moment it germinated in the minds of naturalists, such as Darwin and Wallace, the theory of evolution was bound to mushroom into a scientific, ideological, and political battle.

Evolution was at the center of the war between progress and reaction.

Philosophically, the theory broke with both Plato and Aristotle. It left the Great Chain of Being in pieces. And it destroyed the final justification for Aristotle's idea of purposeful final causes. It is little wonder that many theologians, whose entire worldview was finalistic, renounced evolution with hysterical fervor. The theory might have been accepted long before. But the opposition of clerical and landed interests proved great. They instinctively saw that the victory of evolution might mean the end of any validation of a divine ordering of the world.

Even more scandalous was the idea that man himself, that unique end to creation, was little more than a remarkably successful ape. After Copernicus and Galileo had produced such a potentially Earth-shattering cosmic combination of theory and observation, Newton had come to the rescue. Through his system of the world he had mostly reestablished the credibility of design. So the picture of creation stayed relatively unscathed, especially that of man being in the image of God. After evolution, however, the book of Genesis would be little use as a literal account of history.

In truth, evolution should have completely shattered the doctrine of religion and with it the eternal values of idealist philosophy. But both were to recover only too easily. At first, the theologians thrashed around, frantically seeking a face-saving formulae. Ultimately, they claimed the truth was to be found on another plane, free of contradiction from vulgar facts.

An example of their remarkably silly hypotheses was the idea that God, in his wisdom, had buried the fossils in the rocks to tempt free-thinking geologists into perdition. The notion was the brainchild of English naturalist, and popularizer of natural science, Philip Henry Gosse. Happily, it was thought too far-fetched an explanation to provide valid escape. In the course of the twentieth century, the papal authorities have declared ex cathedra that the first chapter of Genesis must now be understood in an allegorical sense. Save a few fundamentalists, the controversy must be deemed to be over.

In contrast to clerical prejudice, the conception of evolution agreed with the Victorian bourgeoisie. Champions of nineteenth-century science, they found in the theory an affable scientific validation of rampant competition. As a result, it became a weapon in the hands of the materially minded industrialists, especially against sentimental Tories, on the one hand, and dangerous socialists like Wallace, on the other. In their eyes, it gave a scientific blessing to the cause of unfettered commerce and seemed to justify the dripping wealth of the rich by the doctrine of the "survival of the fittest."

The old order was crumbling. And with it the age-old excuse for the dominance of classes or races. But a new mantra dawned. New excuses were needed to justify the conquest and subjugation of inferior by superior peoples. So in many regards, the *Origin of Species* arrived at the perfect time. For the implicit message in the theory of evolution was adopted by the radical, anticlerical wing in politics and economics. It accorded with their views on laisser-faire and self-help. It seemed to vindicate all that was going on in the capitalist world. The ruthless exploitation of the dominions, even war itself, could be warranted by comparison with nature.

The prevailing school of philosophical thought tended to agree, in particular those following John Stuart Mill, the French philosopher Auguste Comte, and the main English sociological theorist of the Victorian era, Herbert Spencer. They too defended, in terms of science and logic, the freedom of private enterprise. They praised the nineteenth century as the era in which man had at last found the right way. True, things were not perfect. There were still some abuses such as slavery to be swept away. But progress would continue. And that progress would be a direct extension of the present: more machines, more inventions, and even more accumulation of wealth.

Wallace was part of a growing movement of artists, poets, and thinkers who protested against the horrors of such excess, against the vulgar flaunting of wealth, suggesting that something was desperately wrong at the very heart of nineteenth-century opulence. Perhaps this is the key difference between Darwin and Wallace. There was political potential in the theory of evolution.

Wallace recognized in evolution the bridge between natural and humane studies. However, evolution's main proponents, who were in positions of power and responsibility, showed a strong reluctance to push home such principles. As Davies points out:

> Wallace, who until late in his life had no inkling of the scandalous background to events at the Linnean Society, had not endeared himself to the establishment by arguing for unpopular causes ... He argued vehemently against the "barbarism" shown towards the Australian Aborigines by English settlers and questioned the "civilising influence" of Christian missionaries in the northern Celebes. He also supported land reform in Britain, arguing that without a more equitable distribution of land, it would not be possible to have a just society. A mixture of snobbery and ridicule diminished his reputation. Darwin's reputation, on the other hand, has increased dramatically as the power of organised religion has declined.[17]

Those such as Wallace realized that, in its emphasis on the kinship of man and the animals, the theory of organic evolution also held the seeds of human liberation. It held implications of history in nature and law in society. But the social evolution of humanity was obscured. Instead, a purely biological message was broadcast. One that in turn would lead to the absurdity of the Nietzschean superman, the rationalizing of race theories, and imperialism.

But there is also a cultural curiosity in the contrast between Darwin and Wallace.

And that is the two-culture split, famously expounded a century later by British scientist and novelist C. P. Snow. In the Rede Lecture in 1959, Snow introduced into the general lexicon a shorthand for the difference between two attitudes, those of the sciences and the humanities. Snow characterized the attitude of *science* as one in which the observer can objectively make unbiased and nonculturally embedded observations about nature. In contrast, he characterized the attitude of the *humanities* as a worldview in which science was seen as embedded within language and culture. Significantly, Snow's lecture was subsequently published as *The Two Cultures and the Scientific Revolution*.

In many ways it was at this time, the days of Darwin and Wallace, that the two-culture split began. Humanists such as Wallace, and the literary and artistic movement, rejected those aspects of science that they understandably felt had identified itself with the machine age and everything in its train. The effect was to prevent any cooperation between the two branches of intellectuals, without which any constructive criticism of the economic and social system was impossible.

With this in mind, we can only imagine the wary skepticism with which the likes of Darwin, Lyell, and Hooker would indulge Wallace. Mainstream scientists were blunted by a quite deliberate turning of their backs on anything—art, poetry, or social justice—that did not come into the ken of their, by now, rather specialised work. Humanists of a more literary bent than Wallace never knew enough of how science worked to have anything other than ineffective emotions about it. Wallace was therefore more dangerous. Had Darwin not warned about "atheistic agitators and social revolutionaries"? Were these not the days when the specter of communism was haunting Europe?

In the two hundred years or so since Galileo, science had been transformed. At the time of the Tuscan, science was but a liberating spirit, seen by only a few choice souls such as Bruno, Bacon, and Galileo himself. In the days of Darwin, science had truly become a material force. It was capable of changing the pattern of life. And its latent power was clear to everyone.

By the second half of the nineteenth century, individual scientists could not really avoid a dilemma. For when they considered the eternal order of nature, they also had to consider the consequences of interfering with nature, using the new forces of science and technology. They were inevitably torn by conflicting impulses. Like Darwin, most were drawn from the middle and upper classes and were associated with the great contemporary movements of capitalism. Even the rare individual recruit from the working classes, such as the English chemist Michael Faraday, was easily assimilated to the cause.

But as scientists they could not help but see the spoil. There was great social damage in the results of their efforts, which were being increasingly used for private enrichment, and were not leading to the improvement of the general lot of man. So much for progress. Only a very few scientists took a conscious part in denouncing such developments. One was English science fiction writer, H. G. Wells. The other was Alfred Russel Wallace.

SCIENCE OR MONKEY BUSINESS

So there was a resonant political dynamic to the development of Darwinism over Wallaceism. But there was also a strong vector of science. Rather than Darwin's radical materialism, Wallaceism was a different kind of evolutionary theory, one immersed in humanism. By 1864, Wallace had published a paper, "*The Origin of Human Races and the Antiquity of Man Deduced from the Theory of 'Natural Selection'*." It was his application of the theory to man. Darwin had not yet communicated the subject in the public domain, although Thomas Huxley, Darwin's "Bulldog," had done so in *Evidence as to Man's Place in Nature*.

Soon afterwards, Wallace was converted to spiritualism. He maintained that natural selection could not explain mathematical, artistic, or musical genius. Nor could it account for metaphysical musings, wit, and humor. For Wallace, "the unseen universe of Spirit" had mediated at least three times in history. First was the generation of life from the primeval soup of the early Earth. Next was the development of consciousness in the higher animals. And last was the maturity of the higher mental faculties in man.

Wallace believed that the rationale of the universe was the development of the human spirit. He simply would not agree that human intellect could be explained by natural selection. A higher spiritual force had to be at work. When Darwin read Wallace's paper outlining this opinion in the April 1869 issue of *The Quarterly Review*, he viewed it as sacrilege and scratched an intemperate "NO!!" in margin of his copy.

The contrast between the rational Darwin and the spiritual Wallace inspires comparison with Galileo and Kepler. Whereas Galileo was wholly and frighteningly modern, Kepler never severed himself from the mystical Middle Ages. Unlike Galileo, who was devoid of any spiritual leanings, Kepler was struck by the magical implications of science. Like Wallace, Kepler's books on science attempted to lay bare the ultimate secrets of the cosmos. They were a hotchpotch of geometry, music, astrology, astronomy, and the occult. And yet, Kepler conjured up the three laws of planetary motion from such a work of lavish fantasy and Wallace the theory of evolution through natural selection.

The paths of Wallace and Darwin diverged. Darwin's materialism held fast. In two books, *The Descent of Man* (1871) and *The Expression of the Emotions in Man and Animals* (1872), he looked at the evolution of human psychology and its continuity with the behavior of animals. Darwin's conclusion was "that man with all his noble qualities, with sympathy which feels for the most debased, with benevolence which extends not only to other men but to the humblest living creature, with his god-like intellect which has penetrated into the movements and constitution of the solar system—with all these exalted powers—Man still bears in his bodily frame the indelible stamp of his lowly origin."[18] The books proved very popular.

Wallace's evolutions, however, continued in a spiritual vein. Wallace developed a different kind of evolutionary theory, a form of teleological evolution. His was an evolution of design and purpose, one that held that all things were designed for, or directed toward, a final result. As Wallace's biographer, Martin Fichman, points out, "Wallace's emerging evolutionary worldview was compatible with a broader spiritual and teleological framework that would become more overt on his return to England,"[19] and that Wallace's observations of orang-utans led him "to invoke more explicitly the concept of design in nature."[20]

Their respective evolutions after the publication of *Origin* help expose the fact that the choice of Darwin or Wallace was embedded in a long struggle, part-scientific, part-political. For the Darwinians, Wallace's particular strain of evolutionism would only have endangered the theory and ruined years of promotion for the superiority of natural selection in the world of science.

And for the radical, anticlerical wing in politics and economics, which adopted an assumed message in the theory of evolution, it is little wonder they chose Darwinism and radical materialism. Given the ongoing war between progress and reaction, they could not risk any flaky association with religion, conventional or unconventional.

What are we to make of Darwin in the context of this aftermath? Like Galileo before him, he seems to have truly been a man of his times, acutely prone to the pervading politics and culture. The conventional portrait painted of Darwin is that of the quiet, mild-mannered, and meticulous naturalist, watchful, orderly, forever thoughtful. As one of his many biographers, Mark Ridley wrote, "Wherever Darwin was—at the dinner table and watching people laugh and have a good time, or visiting the doctor with his children—he was observing carefully, reflecting, questioning and tentatively fitting his ideas into a grand theoretical system."[21] And yet, Darwin also has his decriers. With remarkable similarity to the case of Galileo, this other Darwin is seen as a rather pathetic attention getter, interested more in fame than facts, worried more about reputation than science, a borrower, a poseur, and a cheat.

Part V

The Prestige

CHAPTER 10

THE KUDOS

FAITH IN DOUBT

Penned in exile on the same Danish isle where quantum physicist Niels Bohr worked on his doctorate, Bertolt Brecht's *Life of Galileo* is his greatest play. The original version of the drama was written in 1938, with Brecht being "helped in the reconstruction of the Ptolemaic cosmology by assistants of Niels Bohr who were working on the problem of 'splitting' the atom."[1] The work focussed on the troubled Renaissance physicist, a personal touchstone for those who resist despotic regimes in the name of intellectual freedom. The play's concerns included the conflict between dogma and scientific evidence, as well as the value of steadfastness in the face of oppression.

But the atomic bomb changed everything for Brecht. History's first nuclear weapons exploded over Japan, as Brecht was beginning the second, or "American," version of his great play. "The atomic age made its debut at Hiroshima in the middle of our work. Overnight the biography of the founder of the new system of physics read differently," wrote Brecht.[2]

In the shadow of the atomic bomb, *Life of Galileo* was transformed. What once had been a drama on science as liberator from an irrational worldview was recast. Brecht's new take was a tale of a celebrated scientist who fails to acknowledge his responsibility to humankind, and ends up colluding in the misuse of science. The astronomer's recantation before the Church authorities came to represent the First Fall of Philosophy. "Galileo's crime can be regarded as the 'original sin' of modern natural science,"[3] Brecht wrote. Galileo was no longer a hero, but a traitor.

Brecht's passion for life and science is captured impeccably in *Life of Galileo*, a play that he worked on more than any other. But Cold War audiences would have immediately seen that this was not "just" a history play. *Life of Galileo* is also about politics and the purpose of science. Like Francis Bacon, Brecht believed that science should help liberate, help relieve the drudgery of human existence. "By discrediting the Bible and the Church," wrote Brecht, "these sciences stood for a while at the barricades on behalf of all progress."[4]

After all, asks Brecht, what is the point of the discoveries of Galileo and his fellow physicists if all they ever produce is bigger and better bombs? Post-bomb, Galileo's final speech was rewritten to address the responsibility of scientists, saying, "I hold it that the only proper goal of science is to relieve the miseries of human existence. If scientists, cut off from the masses by selfish rulers, seek merely to heap up knowledge for its own sake, then science is a cripple and your new inventions will merely bring about new drudgeries."[5]

Today the play is as sharply topical as ever. Brecht's complex reconstruction of a seventeenth century Italian astronomer's validation, then recantation, of the Copernican model of the solar system is an enduring parable of human responsibility in the battleground of ideas. It is just as persuasive as a critique of the ideologies and power structures of the twenty-first century.

The most telling lesson of the play relates to what you do with your beliefs, how you implement your ideas. Life today is atomized. It is seemingly polarized into discrete compartments, work and life, reason and emotion, science and culture, the personal and the political. But life and science are indivisible. As Brecht shows, it is just this atomization, this kind of divide and rule that makes people manipulable.

And yet science is no more divisible from politics than humanity is from a life of the senses. In the first version of the play, Galileo was a flawed hero to science. True, he recants. But he redeems his betrayal by secretly copying his long-awaited *Discorsi*, which is smuggled out, gifting a ray of truth to a darkening world. In the American version, with a typically dialectical Brechtian twist, the events are the same. But now Galileo fails his greatest test. *The science is less important than what it stands for.*

To signify a world in constant change, Brecht's choice of Galileo and the telescope is sublime. The manufacture of the instrument itself, the weapon of Galileo's discoveries for science, embodies a time when long familiar theory had finally been married to practice. In that century of revolutions, there was a crucial moment, thought Brecht, when ordinary people were ready to rally to reason and spin the spyglass on to the activities of "their tormentors: the princes, landlords and priests."[6]

In Chapter 8 the idea was introduced of Copernican astronomy holding a latent potential to liberate Earth as well as heaven. The republican radicals of the English Revolution, for example, saw in the new science the possibility of gifting a rational approach to all aspects of human life, to democratize all things, to teach science and politics in every parish by an elected nonspecialist. And to eradicate the difference between experts and novices.

Brecht agreed. In Galileo he identified a man who had taken science away from the Latin Bible and its black-coated ministers and into the streets and marketplaces. At that moment Brecht felt Galileo personified the great potential. To call the bluff of the papal authorities, Galileo could have slammed shut the yawning gap between the scholar and the layman, delivered science out of the hands of the specialized few mumbo-jumbo men, and begin the work toward a far wider spread of knowledge.

But knowledge remained a monopoly. Galileo gave in to the threats of torture. "Had I stood firm," says Galileo in Brecht's play, "the scientists could have developed something like the doctors' Hippocratic oath, a vow to use their knowledge exclusively for mankind's benefit. As things are, the best that can be hoped for is a race of inventive dwarfs, who can be hired for any purpose."[7] Science for science's sake is a terrible trap. Its aim is not "to open the door to infinite knowledge but to put a limit to infinite error."[8] One day, he predicts, the gap between science and mankind will yawn so wide that "your cry of triumph at some new discovery will be echoed by a universal cry of horror."[9]

It was the last thing the radicals wanted—Galileo's recantation. As Brecht suggested in one of the forewords to his play, "It is true that a swing-back took place in the following centuries, and these sciences were involved in it, but it was in fact a swing instead of a revolution; the scandal, so to speak, degenerated into a dispute between experts. The Church, and with it all the forces of reaction, was able to bring off an organised retreat and more or less reassert its power. As far as these particular sciences are concerned, they never again regained their position in society, neither did they ever again come into such close contact with the people."[10]

Brecht's skepticism is in the best spirit of science itself. He has a healthily ambivalent attitude to authority, which greatly lends Galileo's legend a continued significance. "Already in the original version," he explains, "the Church was portrayed as a secular authority, its ideology as, fundamentally, interchangeable with many others."[11]

Brecht was also under no illusion about human frailty and the skill with sophistry and cajolery that keeps the powerful in place. The legend shows how little it takes to subvert even those of the goodwill, drive, and determination of Galileo. And how bold the powerful are in support of their privilege. At one stage, Barberini, the new pope, and a mathematician, notes irritably that it is impossible to sanction the use of Galileo's star charts, as sailors are demanding, while condemning the theory upon which they are based. "Why not?" replies an Inquisitor. "One cannot do otherwise."[12]

For the great playwright, skepticism and doubt are the antidote. Disbelief can move mountains, he says elsewhere, and the pleasure and necessity of doubt is a reoccurring theme throughout the play. The role of knowledge is to turn us all into doubters. It is a suitably scientific message for what has often been called an "optimistic tragedy." It is worth remembering, in today's age of fundamentalisms, when you read of the need for great leaders in politics or in science. "Unhappy the land that has no heroes!" says a believer bitterly when Galileo recants. "No," says Galileo, "unhappy the land that needs heroes."[13]

GALILEO AND DARWIN AS HEROES

To many people, Galileo and Darwin are great heroes. Brecht suggested, for example, that "from the first, the key-stone of the gigantic figure of Galileo was his conception of science for the people. For hundreds of years and throughout the whole of Europe, people had paid him the honour, in the Galileo legend, of not believing in his recantation."[14] And reading a conventional view of the scientific revolutions associated with Galileo and Darwin, it is easy to get the impression that progress in their respective disciplines was due solely to the genius of these great men. Each allegedly bestowed upon the world their momentous and revelatory insight into the secrets of nature.

But the more detailed study of their myth has revealed an underlying truth. As has been seen through a longer study of the history of both cases, for Galileo with space and Darwin with time, the ancient Greeks laid down significant foundations to each discipline. For centuries afterward, depending strongly on varying conditions of culture, society, and economy, many more ordinary thinkers and workers in these respective fields gifted to science a gradual advance of tradition and custom.

Providentially, Galileo and Darwin lived at times of radical movements in science and society. Both the 1600s and 1800s were centuries of revolution in Europe. The seventeenth century saw not only the beginning of European colonization of the Americas, the Thirty Years' War, the Dutch Revolt, and the English Revolution, but also the start of the scientific revolution, a movement to which Galileo belonged. The nineteenth century of Darwin witnessed the continuation of the revolution in France. It saw the Spanish, Portuguese, Chinese, and Holy Roman empires crumble, and the rise of the British Empire, which controlled one quarter of the world's population and one third of the land area.

Both Galileo and Darwin were involved with innovatory leaps in science that led to decisive changes in the way we look at the world, and man in it. By associating with these radical movements, Galileo and Darwin were able to make significant contributions to the social development of science. And when this contribution is studied in its contemporary social setting, we realize they are truly products of their time.

So, a more detailed account of the context of their personal contributions is quite revealing. For not only does it show the way in which they were subject to the same sway of social influence, the same sorry compulsions. It also shows how both Galileo and Darwin were creatures of the culture in which they swam, how immersed they were in the milieu of their days. And only by seizing the moment chanced by their times were they able to make that innovatory leap that led to such legendary lives, such celebrated contributions.

And yet Galileo and Darwin have their decriers. The growing legend of each man has also sown considerable seeds of doubt in the minds of a number of scholars and writers. Both are seen as pathetic attention seekers, looking for fame more than facts. Both are painted as far more concerned about reputation than science. And both in their mission to seek fame and fortune have gained along the way a certain notoriety to beg, borrow or steal to get there.

Controversy hounded Galileo. And that controversy was not only with the papal authorities. There are those for whom the legend of Galileo makes him a martyr for science. There are those for whom his legend is based on inventions he never made, such as the telescope, microscope, or pendulum clock, and theories he never instigated, such as the law of inertia, or the Sun-centred solar system.

In fact, through his own ignorance of Kepler's work, he was unable to prove the heresy of Copernicanism. In fact, he was never tortured by agents of the Inquisition, nor languished in its dungeons. In fact, many of the Galilean disputes were over science, not faith. The controversy over the precedence of the discovery of sunspots, for example, shows a man of "misplaced sense of self-importance"[15] and the "cold, sarcastic presumption"[16] by which he "managed to spoil his case throughout his life."[17] And if the legend of revolutionary astronomy of the time has an auto de fé, the charred remains at its burning stake belong to Giordano Bruno, not Galileo.

Darwin is an entirely different animal, but still the similarities are striking. He may not have been "loud, forceful, argumentative and combative" like Galileo.[18] Indeed, as we have seen, Darwin's reputation is that of the quiet, unassuming naturalist. But scholars, such as professors Cyril Dean Darlington and Loren Eisley and writer Roy Davies, paint another picture. They portray and demonstrate a duplicity in Darwin, which sits very uncomfortably with the myth of the revered naturalist.

Darwin's case of priority with natural selection bears a great resemblance to the disputes in which Galileo became embroiled. But Darwin was far subtler in his disputations, and far more calculating. He was a different species of scientist to Galileo. Along with influential friends and colleagues, such as Joseph Hooker and Charles Lyell, Darwin was something of a chimera, a hybrid creature of aristocrat and new "professional" scientist.

Both Darwin and Galileo rested their reputations on the scientific establishments of the day. With Galileo, it was the elite society of the Lincean Academy, whose founding president, the young nobleman Federico Cesi, was the second marquis of Monticelli, soon to become a prince. With Darwin, it was the Linnean Society, the world's premier society of natural history, whose prominent members Hooker and Lyell shepherded him through the thorny accusations of plagiarism after receiving the scoop from Wallace.

For Galileo and for Darwin, their position as "revolutionary" scientists was very much aided by powerful backers from the commanding classes.

The new astronomy deeply interested the rising bourgeoisie, and the discipline gave great impetus to the revolutionary social current of the time. Galileo's disputes, with backers such as Cesi and the Duke of Tuscany, and especially when compared to the lowly Bruno, speak highly of class interest. Galileo's links with the Medici family are significant in explaining the way in which the Church dealt with him relatively leniently. The Medici were the wealthiest family in Europe and counted in their number not only three popes, but also the most famous religious works of Renaissance art. As usurers running one of the most prosperous and respected banks, they wanted to ensure their place in heaven by providing financial backing for the Church.

By the days of Darwin, science had become a material force. Its champions were no longer emerging, but safe, secure, and running the workshops of the world. The new biology also had to deal with clerical prejudice. But by now, the fight was almost over. And the *Origin of Species* arrived at the perfect time, providing a scientific blessing to the cause of unfettered commerce.

WEAPONS OF DISCOVERY

From this rather lofty perspective, it is all too easy to see Galileo and Darwin as "poster boys" for science, thrust forward as pawns or protagonists in a much bigger game.[19] But in many senses, this is quite true. If we were to ask what kind of world, what kind of society effectively bred the weapons of discovery wielded by both men, the economy is quite key. For, echoing Brecht's words in the *Life of Galileo*, *the science is less important than what it actually stands for.*

There is much to be said for the ship as a weapon of discovery. In the very first scene of *Life of Galileo*, Brecht has the Tuscan astronomer declare in his "humble study in Padua":

> For two thousand years men have believed that the Sun and the stars of heaven revolve around them ... The cities are narrow and so are men's minds. Superstition and plague. But now we say: because it is so, it will not remain so. I like to think that it all began with ships. Ever since men could remember they crept only along the coasts; then suddenly they left the coasts and sped straight out across the seas.
> On our old continent a rumour started: there are new continents! And since our ships have been sailing to them the word has gone round all the laughing continents that the vast, dreaded ocean is just a little pond. And a great desire has arisen to fathom the causes of all things ... For where belief has prevailed for a thousand years, doubt now prevails. All the world says: yes, that's written in books but now let us see for ourselves.[20]

In the period between Galileo and Darwin, the agenda for science, warfare, and trade became increasingly blurred. The period of great imperial expansion between the lives of these two scientists used science to drive and justify imperial ambition. The voyages of discovery were linked inextricably with maritime navigation based on astronomical knowledge. For the way in which empires were able to exert their global control relied on knowledge of their territory, knowing where you were, knowing what you owned. That is why astronomy, navigation, and mapping were so important to trade.

The ship was of prime importance. Since all the effective trade routes were wet, commercial control very much depended on the speed and reliability with which long-range voyages took place along those trade routes. The discovery of the Americas and the New World was the impetus for a more aggressive approach to nature. The quest for longitude and dominion in that new world promoted instrumentalism in science. And one of those first instruments or weapons of discovery was the ship itself.

In a very real way, Galileo's telescope was a kind of ship. Witness the way in which he used it. He took himself as observer, and his contemporary world as onlookers, to a place almost no one save Bruno and Kepler had ever imagined. But, like the destinations of a ship, the destinations of the telescope are also public knowledge. If you did not believe what Galileo said in *The Starry Messenger*, you could take a look for yourself.

With the gathering decades came treaties on the new philosophy, such as Francis Bacon's *New Atlantis* and Robert Hooke's *Micrographia*, which, like Brecht, used powerful maritime analogies in their treatise on the new science. The first science fiction stories of the age were narrative explorations of ship voyages in space. Explorations of alien worlds began with journeys to the Moon, such as Kepler's *Somnium* and Francis Godwin's *The Man in the Moone*, and later morphed into the planetary novel with further flights of the fantastic, such as Cyrano de Bergerac's *L'autre Monde* (*Other Worlds*).

Great expeditions were sent out from the centers of empire in London and Paris. James Cook's voyage to measure the transit of Venus led to the discovery of new worlds in the antipodes. And on the eve of the voyage of the *Beagle* came the voyages of the German naturalist and explorer Alexander von Humboldt. Through precision measurement in his extensive travels through South America between 1799 and 1804, Humboldt showed the potential for a new science of the laws governing life on planet Earth. The journals of his voyages proved massively popular in Europe, especially in Darwin's England.

In the post-Napoleonic period and the days of Darwin, there was an obsession with individual freedom and private enterprise, which not only fed into natural selection as an idea, but also meant lower taxes for the rich and wealthy. Science became a passion for gentlemen such as Hooker, Lyell, and Darwin, who found himself on the *Beagle*. The main purpose of the expedition was a hydrographic survey of the coasts of the southern part of South America, producing charts for naval war or commerce.

This brief consideration of the ship itself as a weapon of discovery again places the position of both Galileo and Darwin in perspective. Conventional accounts might picture the gallant Galileo as the modern scientist, or portray Darwin as the last gentleman explorer. The bigger picture blurs both depictions. And their work is set in context against a background of the movement of capital and continents.

THE LONG VIEW

The year 2009 was a cause célèbre, a watershed for the weapons of discovery. Labeled International Year of Astronomy, because it marked the 400th anniversary of Galileo's first use of the telescope, and Darwin200, signaling not only Darwin's birthday, but also the 150th anniversary of the theory of natural selection, the world's scientific press clamored about the question of which "year" mattered most.

The celebrations should have afforded a superb opportunity to probe deeper into history, take the long view, the evolutionary view. What was delivered, of course, was more of the same. Rather than ask acute questions about the conditions of culture and economy that led to such remarkable histories, much of the coverage probed little deeper than the mere surface.

Instead of looking at how political bias and unconscious prejudice played a part in the creativity of both men, one leading science publication symptomatically asked, "Darwin or Galileo: Who Did Most to Cut us Down to Size?"[21]. "A host of thinkers," the publication boasted, "decide who really deserves 2009's anniversary crown."[22] So not only did we get the "great men" myth. We also got Galileo and Darwin pitted against one another in an apparently philosophical head-to-head: "Who has done more to knock humanity off its pedestal?"[23]

It gets worse. Leading scientists, asked for their opinion on such a trifling matter, were quite exposed in their responses. The U.S. theoretical physicist and cosmologist Lawrence Krauss plumped for Galileo. It was clear to Krauss that "anyone who was looking could have seen that humans were animals."[24] Well, apparently not. Not until the nineteenth century and later did a satisfactory science of life on Earth emerge. Krauss's championing of Galileo is due to the Italian having apparently single-handedly, "removed us from the centre of the universe" and having "replaced divinely revealed knowledge with empirical knowledge. Go Galileo!"[25] Some task, for a single Renaissance astronomer.

The situation got no better when they quizzed Matt Ridley, a writer on evolutionary biology. Asked whether Galileo was the "winner" as the man "who demonstrated beyond doubt that the Earth is not at the centre of anything,"[26] Ridley's response was, "Who cares which ball of rock goes round which?"[27] Both opinions beggar belief. The reader is torn between which analysis is the more asinine. Krauss, for believing that the scientific revolution was carried out by a single man, or Ridley, for believing the revolution was overrated in the first place.

The editorial team came to the cursory conclusion that "if it were left to Darwin and Galileo to argue their supremacy, there is no doubt that Galileo would come out on top."[28] It is rather ironic that a revolution so associated with the promotion of doubt as a weapon of discovery should now be identified with the removal of all doubt. Galileo, we are told, was "a great polemicist" who "spent much of his energy on vigorous self-justification." Darwin, on the other hand, "shrank from controversy, leaving others to argue his case."[29]

It is an understandable conclusion, if you look at Galileo and Darwin in superficial isolation. But when on earth did they ever act in isolation? When did they ever act as pure individuals? In each case, for each weapon of discovery, we can at the very least identify an allegiance of instigator, protagonist, and patron, and a collusion of conspirators that helped wage war on convention.

Simply speaking, Copernicus was the instigator, during the Renaissance period, of the Sun-centred solar system. Galileo was its flawed protagonist, the Duke of Tuscany his patron, the Lincean Academy his coconspirators. The telescope was his weapon of discovery. Again simply speaking, Wallace was the instigator, during the nineteenth century, of the theory of evolution by natural selection, another weapon of discovery. Darwin and his "bulldog" T. H. Huxley were its protagonists, and Linnean Society members such as Hooker and Lyell its coconspirators.

But even this version of events is woefully simplistic, of course. The true picture of these "revolutionary" scientists sets them on an even greater stage. Taking the long view has allowed us a chronicle that has swept continents and centuries. The weapons have played their part in upending kings and cosmologies, helped old orders and empires crumble, and almost did to death the dogma of religion and the dark age of faith.

This sense of an evolving knowledge, provided by the long view, allows us to see the appalling cost of continuing to ignore history. For with it, as the "Darwin versus Galileo" example testifies, any intelligent understanding of the place of science in society is lost. Armed with knowledge of science's past, the student of science can better understand the grand ongoing drama of the use and abuse of science.

It is worth considering the Darwin versus Galileo piece from a media perspective as well. Science writers and journalists have to compete in an industry that has "news values." And one of the dominant tendencies in stories about science, even one which is two hundred or four hundred years old, is all about elitism and personalization.[30] Famous scientists are interesting, even minor details of their lives can make headlines. For those with legends attached, such as Galileo and Darwin, this is doubly so. And yet this need not be the focus. Such science as this also deals spectacularly with the new and the strange, which would just as easily sell a news story.

The Darwin versus Galileo piece is a case in point. It forms part of a rather unadventurous way of thinking about the nature of science, symptomatic of a rather ahistorical philosophy. Another facet of the same approach is to believe that current knowledge is the best available wisdom on science. That it has somehow replaced and supplanted all preceding scientific knowledge. According to this belief, current knowledge, too, will become obsolete, displaced by future facts. All useful previous knowledge is subsumed by that of the present; the mistakes of the ignorant, consigned to the dustbin of history. In short, and in the words of American industrialist Henry Ford, "History is bunk."

It is rather ironic, this ahistorical tendency. Especially when you consider the case studies of Galileo and Darwin. For astronomy and evolutionary biology are *historical* sciences. The crucial significance of the theory of evolution was that it introduced a historical dynamic into the field of science. It broke with the orthodox branch of the Greek tradition. It broke with the eternal truths and fixed species of Plato and Aristotle. And it returned to the earlier and heretical branch of the old Ionian philosophers, and of the Atomists, with their focus on rational development and change.

The rise of evolution injected the lifeblood of history into science. "He who … does not admit how vast have been the past periods of time may at once close this volume,"[31] Darwin wrote in the *Origin*. For species to have evolved, the genuine extent of the Earth's past had to be much longer than the six thousand years or so suggested by the Bible. While biology and geology implied the Earth was ancient, they did not, however, prove it.

But once again, *the science is less important than what it stands for.*

The spirit of evolution was gifted to the physicists. They, too, approached the question of the age of the Earth and the age of the

Sun. First they used thermodynamics, that branch of physics concerned with the dynamics of heat energy. Then, late in the nineteenth century, the nuclear age dawned. Radiometric dating, the technique for dating materials using naturally occurring radioactive isotopes, provided age-dating to fields as diverse as geology, astrophysics and cosmology.

By the 1920s, it was becoming clear to astronomers and geologists that the Earth was billions of years old. Rocks that were brought back by Apollo astronauts from the Moon, that natural satellite Galileo had spied through his far-seer centuries before, were dated at around 4.6 billion years old. And a consideration of nuclear matter in motion led the astrophysicists to the conclusion that the Sun, which Galileo had observed to have spots and impurities, is a normal star, about halfway through its 10 billion year evolution.

The Atomists were right all along. Their philosophy of evolutionary change proved key to the cosmos. In the last century or so, biological evolution begat the idea of evolution in science per se. History was revealed by evolution to be not only central, but *big*. Once we realized that the Earth was old, and that all had been fashioned by lengthy processes of change, we began to see that the story of our planet was part of an even older tale. And the materialist bedrock of evolution has influenced our perception of all things: culture, language, the society in which we live, and every single discipline of science itself.

Everywhere we look in nature, we see systems of matter in motion, evolutionary change. The electron clouds that constantly swarm the heart of the atom also make up the hydrogen and helium that is brewed by a star's interior into heavier elements, gifting to each galaxy an evolving chemistry. The swirling disc of gas and dust in the early solar nebula gave birth to our solar system, the planets growing out of the nebula, evolving as they circled the young Sun.

The universe itself is also changing, as is our rapidly evolving conception of it. When Einstein had composed his general theory of gravitation, he had assumed that the universe was static. But before long, telescopic redshift evidence from the likes of Edwin Hubble told the telltale signs of an expanding and evolving spacetime. We have since found out we live in a Universe that would make even Galileo wonder. The small, static, and Earth-centred cosmos he helped demolish has given way to the notion of an evolving Universe so large that light from its outer reaches takes longer than twice the age of the Earth to reach our telescopes.

The Atomists believed in the boundless energy of a world in flux, and man's power to alter it, by learning its laws. Likewise, one of the aims of modern science is to explain the sweep of phenomenal changes in Earth and Sky over the vastness of deep space and time. Ever since the scientific revolution, science has sought not just to explore, but to exploit nature. To master it.

It will not be an easy task. Research in biology has revealed that evolution occurs both above and below the conventional Darwinian level of the adaptive struggle of species for survival. Below the conventional level is the random mutation of genetic material leading to unsolicited and serendipitous change. And above the conventional level is the prospect of cosmic impact catastrophes, which have punctuated life's pathway on Earth, and are used to mark geological time itself. Since we are prey to all three levels of evolutionary change, there is little left of man's God-given right to our assumed dominion over nature. And physicists believe that even the very laws of nature themselves seem to have evolved from simpler, original laws.

THE SKEPTICAL ENQUIRER

The pursuit of science requires particular courage. It is concerned with knowledge of the world, acquired through discovery and doubt. Making this knowledge of the world available for everyone, as did Galileo and Darwin, science strives to make skeptics of us all. And yet conventional accounts of science's history are often little more than a mantra, preached to the greater part of the population. The intention often seems to keep readers in a nacreous haze about history, telling the tale of science as the preserve of the exceptional genius and obscuring the true machinations of society and culture.

But we should be skeptics of science too, and skeptical about the histories we are given. It is the best antidote to the vulgar and vigorous tendency to paint the past in stark monotones of heroism and villainy. A closer study of the evolutions of Galileo and Darwin reveals that the fight over the measurability of the heavens and the battle for a science of life on Earth was won through doubt. It also reveals that science makes its biggest strides forward when the gap between science and laity is narrowest. When the device of doubt delights the great public as well as the scientists themselves.

Society today, in its material features, would be impossible without science. Indeed, many contemporary intellectual and moral aspects of civilization are also deeply influenced by science. The dissemination of scientific ideas has been the most crucial dynamic in the shaping of contemporary thought. But society is also faced with great fears. Annihilation through weapons of mass destruction, concerns over climate change, and a mushrooming global population sit alongside the hope of longer and better lives through the application of medical science.

Progress in science seems to bring pitfalls as well as promise. The way science manifests itself in society, its freedoms and secrecies, its use in education as well as in war, its relationship to governments and culture, all this presents us with a host of great challenges. How are such questions to be addressed, if at all?

A working critical understanding of the relationship between science and society is essential. One that requires knowledge of the *history* of science and of society. In science, more than in any other human endeavor, it is vital to look to the past in order to understand the now and to plan the future.

We need to understand science. Not just the technical detail, but how science is the catalyst by which our societies are being swiftly changed. And to understand how, we need not only look to the present. We must also look to the past, at how science evolved, how science has responded to the successive forms of society, and how it has helped shape them. Of course, scientists do not run society. They are not ultimately responsible for the troubles of our times. They do not have some kind of operational power over civilization. But the use of science is not totally out of our hands.

NOTES

CHAPTER 1

1. J. D. Bernal, *Science in History*, vol. I (London, 1965).
2. Jean Schneider, *Interactive Extra-solar Planets Catalog. The Extrasolar Planets Encyclopedia*, http://exoplanet.eu/catalog.php (accessed March 26, 2009).
3. Bernal, *Science in History*.
4. *Encyclopedia Britannica*. 1955 ed., II-582c.
5. Ibid., II-582d
6. Bernal, *Science in History*.
7. A. Koestler, *The Sleepwalkers: A History of Man's Changing Vision of the Universe* (London, 1959).
8. Book III, Herodotus Collected Works, chap. 13. Quoted by Ch. Seltman, "Pythagoras," in *History Today*, August 1956, 7–11.
9. Proclus, quoted in T. Danzig, *Number, The Language of Science* (London, 1942), 101.
10. B. Farrington, *Greek Science* (London, 1953).
11. Bernal, *Science in History*.

CHAPTER 2

1. J. D., Bernal, *Science in History*, vol. I (London, 1965).
2. See about painting at start of next chapter.
3. Bernal, *Science in History*.
4. Ibid.
5. K. S. Guthke, *The Last Frontier: Imagining Other Worlds from the Copernican Revolution to Modern Science Fiction* (New York, 1990).
6. Quoted in Bernal, *Science in History*.
7. Censorinus, *De Die Natali*, IV, 7.
8. S. F. Mason, *A History of the Sciences* (London, 1953).
9. H. A. Diels, *The Fragments of the PreSocratics* (Berlin, 1952), B54.
10. Bernal, *Science in History*.
11. Diogenes Laërtius, Book III, 20.53.
12. J. Levenson, *Circa 1492: Art in the Age of Exploration* (New Haven, 1991).
13. Cicero, *de Finibus*, vol. 19; Strabo, xvi.
14. Clement of Alexandria, *Stromata*, i.

15. Diogenes Laërtius, Book IX, 72.
16. Simplicius, "Commentary on Aristotle's Physics," 1121, 5–9.
17. J. Stevenson, *The Complete Idiot's Guide to Philosophy* (Indianapolis, 2005).
18. L. L. Whyte, *Essay on Atomism: from Democritus to 1960* (London, 1961).
19. Bernal, *Science in History*.

CHAPTER 3

1. H. W. Janson, and A. F. Janson, *History of Art: The Western Tradition,* 7/E (New Jersey, 2007).
2. A. N. Whitehead, *Science and the Modern World* (Cambridge, UK, 1953).
3. A. Koestler, *The Sleepwalkers: A History of Man's Changing Vision of the Universe* (London, 1959).
4. Ibid.
5. Ibid.
6. Ibid.
7. P. Duhem, *Le Système du Monde: Histoire des Doctrines Cosmologiques de Plato à Copernic* (Paris, 1917).
8. Koestler, *The Sleepwalkers*.
9. Ibid.
10. Duhem, *Le Système du Monde*.
11. J. L. E. Dreyer, *History of Planetary Systems from Thales to Kepler* (Cambridge, UK, 1906).
12. T. L. Heath, "De facie in orbe lunae," Chap. 6 in *Greek Astronomy* (London, 1932).
13. J. D. Bernal, *Science in History,* vol. I (London, 1965).
14. Ibid.
15. Ibid.
16. Ibid.
17. H. Dingle, *The Scientific Adventure* (London, 1952).
18. B. Russell, *Unpopular Essays* (London, 1950).
19. Bernal, *Science in History*.
20. *The Republic of Plato*, book VII, trans. Thomas Taylor (London, 1804).
21. G. B. Grundy, article on "Greece," Encyclopedia Britannica, x-780c (1956).
22. B. Farrington, *Greek Science* (London, 1953).
23. Ibid.

CHAPTER 4

1. M. Brake and N. Hook, *Different Engines: How Science Drives Fiction and Fiction Drives Science* (London, 2007).

2. T. Kuhn, *The Copernican Revolution: Planetary Astronomy in the Development of Western Thought* (Harvard, 1957).
3. C. Grandgent, *Discourses on Dante* (Harvard, 1924).
4. Brake and Hook, *Different Engines.*
5. J. D. Bernal, *Science in History,* vol. I (London, 1965).
6. St. Augustine, *Works,* ed. Marcus Dods (Edinburgh, 1870).
7. Bernal, *Science in History.*
8. A. C. Crombie, *Robert Grosseteste* (Oxford, 1953).
9. T. Kuhn, *The Copernican Revolution: Planetary Astronomy in the Development of Western Thought* (Harvard, 1957).
10. Ibid.
11. J. Milton, *Paradise Lost,* 2nd ed. (London, 1674).
12. A. Koestler, *The Sleepwalkers* (Penguin, 1959).
13. A. C. Crombie, *From Augustine to Galileo* (London, 1952).
14. Crombie, *Robert Grosseteste.*
15. Bernal, *Science in History.*
16. E. G. R. Taylor, *Late Tudor and Early Stuart Geography 1583–1650* (London, 1954).
17. Brake and Hook, *Different Engines.*

CHAPTER 5

1. M. Brake and N. Hook, *Different Engines: How Science Drives Fiction and Fiction Drives Science* (London, 2007).
2. A. Lovejoy, *The Great Chain of Being* (Cambridge, MA, 1936).
3. *De animalibus historia,* quoted in ibid.
4. St. Thomas Aquinas, *Summa contra Gentiles,* Trans. Joseph Rickaby (London, 1905).
5. Lovejoy, *The Great Chain of Being.*
6. M. de Montaigne, *The Complete Essays,* trans. M. A. Screech (1987) (London, 1595).
7. A. Pope, *An Essay on Man* (London, 1734).
8. Quoted in E. M. W. Tillyard, *The Elizabethan World Picture* (London, 1943).
9. *Groundhog Day* is a 1993 comedy movie, directed by Harold Ramis, in which an egocentric Pittsburgh TV weatherman who, during a hated assignment covering the annual Groundhog Day event (February 2) in Punxsutawney, finds himself repeating the same day over and over again.
10. S. Sambursky, "The Stoic Doctrine of Eternal Recurrence," in *The Concepts of Space and Time,* ed. M. Capek (Boston, 1976).
11. The Seven Sages (of Greece) was the title given by ancient Greek tradition to seven early sixth century BC philosophers, statesmen, and lawgivers who were renowned in the following centuries for their wisdom. Along with Thales of Miletus, the other six were Pittacus

of Mytilene, Bias of Priene, Solon of Athens, Cleobulus of Lindus, Myson of Chen, and Chilon of Sparta.

12. J. D. North, "Chronology and the Age of the World," in *Cosmology, History and Theology*, ed. W. Yourgrau and A. D. Breck (New York, 1977).
13. Quoted in Brake and Hook, *Different Engines*.
14. C. Ogburn, *The Forging of Our Continent* (New York, 1968).
15. Lovejoy, *The Great Chain of Being*.
16. L. Eisley, *The Firmament of Time* (New York, 1970).
17. Lovejoy, *The Great Chain of Being*.

CHAPTER 6

1. The Duomo, the Basilica di Santa Maria del Fiore, is the cathedral church of Florence.
2. M. Brake and N. Hook, *Different Engines: How Science Drives Fiction and Fiction Drives Science* (London, 2007).
3. J. D. Bernal, *Science in History*, vol. I (London, 1965).
4. Brake and Hook, *Different Engines*.
5. C. S. Sherrington, *The Endeavours of Jean Fernel* (Cambridge, UK, 1946).
6. J. Burckhardt, *The Civilisation of the Renaissance in Italy* (London, 1944).
7. B. Brecht, *Leben Des Galilei*, trans. M. Wilson (Wales, 1938).
8. Bernal, *Science in History*.
9. E. G. R. Taylor, *Late Tudor and Early Stuart Geography, 1583–1650* (London, 1934).
10. H. Butterfield, *The Origins of Modern Science* (London, 1957).
11. Royal Astronomical Society, "Nicholas Copernicus, De Revolutionibus, Preface and Book I," trans. J. P. Dobson and S. Brodetsky, *Occasional Notes*, no. 10 (1947).
12. Ibid.
13. A. Koyré, *La Révolution Astronomique* (Paris, 1961); T. S. Kuhn, *The Copernican Revolution* (Harvard, 1957); and A. Koestler, *The Sleepwalkers* (London, 1959).
14. Royal Astronomical Society, *Nicolaus Copernicus*.
15. S. J. Dick, *Life on Other Worlds* (Cambridge, UK, 1998).
16. Bernal, *Science in History*.
17. A. D. Weiner, "Expelling the Beast: Bruno's Adventures in England," in *Modern Philology* 78, no. 1 (1980): 1–13.
18. S. J. Dick, *The Biological Universe* (Cambridge, UK, 1996).
19. R. G. Ingersoll, *The Collected Works of Robert G. Ingersoll* (New York, 1902).
20. "Il Sommario del Processo di Giordano Bruno, con appendice di Documenti sull'eresia e l'inquisizione a Modena nel secolo XVI," ed. Angelo Mercati, in *Studi e Testi*, vol. 101 (Vatican, 1952).
21. G. W. Foote and A. D. McLaren, *Infidel Death-beds* (London, 1886).

22. Ibid.
23. M. Rose, *Alien Encounters: Anatomy of Science Fiction*, (Harvard, 1982).
24. J. J. Kessler, *Giordano Bruno* (Stark, 1997).
25. S. Rabin, "Nicholaus Copernicus," in *Stanford Encyclopaedia of Philosophy* (2005); and W. Turner, "Giordano Bruno," in *Catholic Encyclopaedia* (1908).
26. *Summary Of The Trial Against Giordano Bruno*, http://asv.vatican.va/en/doc/1597.htm (accessed May 3, 2009).
27. Ingersoll, *The Collected Works of Robert G. Ingersoll*.
28. A. van Helden, "The Invention of the Telescope," in *Transactions of the American Philosophical Society*, 67, no. 4 (1977).
29. H. G. Topdemir, *Takiyüddin'in Optik Kitabi* (Ankara, 1999).
30. G. della Porta, *Natural Magick* (Sioux Falls, 2005).
31. Ms. F, 25r., Institut de France, Paris.
32. Ms. A, 12v., Institut de France, Paris.
33. K. von Gebler, *Galileo Galilei and the Roman Curia* (London, 1879).
34. G. Galilei, *Siderius Nuncius* (Venice, 1610).
35. George Fugger in a letter to Johannes Kepler, *G.W.* XVI (16 April 1610): 302.
36. G. Galilei, *Siderius Nuncius* (Venice, 1610).
37. Ibid.
38. Ibid.
39. Ibid.
40. Ibid.
41. Ibid.
42. Ibid.
43. Brake and Hook, *Different Engines*.
44. G. Galilei, *The Starry Messenger*, trans. Albert Van Helden (Chicago, 1610).
45. J. Donne, *Ignatius his conclave: or his inthronisation in a late election in hell* (1611).
46. Logan Pearsall Smith, ed., *Life and Letters of Sir Henry Wotton* (Oxford, 1907).
47. Ibid.
48. Ibid.
49. Ibid.
50. Ibid.
51. Ibid.
52. J. Milton, *Paradise Lost* (1667).

CHAPTER 7

1. L. Eisley, "Charles Lyell," *Scientific American* 201 (1959): 98–106.
2. S. Toulmin, *The Discovery of Time* (Chicago, 1982).
3. Eisley, *Charles Lyell*.

4. J. W. Knedler, *Masterworks of Science* (New York, 1973).
5. Eisley, *Charles Lyell*.
6. Ibid.
7. W. W. Bartley, "What Was Wrong With Darwin," *New York Review of Books* (September 15, 1977): 37.
8. C. Darwin, *The Voyage of the Beagle* (London, 1839).
9. F. Darwin, *Charles Darwin's Autobiography* (New York, 1950).
10. Knedler, *Masterworks of Science*.
11. Ibid.
12. Ibid.
13. Ibid.
14. B. Aldiss, *Trillion Year Spree* (London, 1986).
15. Ibid.
16. Ibid.
17. In *Encyclopaedia Britannica*, 15th edition, vol 5, 492.
18. R. D. Keynes, *The Beagle Record* (London, 1979).
19. E. Darwin, *Zoonomia*, vol. 1 (London, 1818), 397, 400.
20. M. Ruse, *The Darwinian Revolution* (Chicago, 1979).
21. C. Darwin, *The Origin of Species by Means of Natural Selection* (New York, 1872).
22. Ibid.
23. Ibid.
24. Ibid.
25. Bartley, *What was Wrong with Darwin*.
26. J. Carey, *The Faber Book of Science* (London, 1995).

Chapter 8

1. J. Milton, *Areopagitica: A speech of Mr John Milton for the liberty of unlicensed printing to the Parliament of England* (London, 1644).
2. Galileo, *Le Opere di Galileo Galilei*, vol. XI (Florence, 1968).
3. F. Sizzi, *Dianoia Astronomica, Optica, Physica* (Venice, 1611).
4. S. Drake, *Discoveries and Opinions of Galileo* (New York, 1957).
5. M. Caspar, *Johannes Kepler* (Stuttgart, 1948).
6. Ibid.
7. Drake, *Discoveries and Opinions of Galileo*.
8. Ibid.
9. Ibid.
10. Ibid.
11. M. Brake, *On the Plurality of Inhabited Worlds: A Brief History of Extraterrestrialism* (Wales, 2003).
12. P. Parrinder, *Learning from Other Worlds: Estrangement, Cognition, and the Politics of Science Fiction and Utopia* (Liverpool, 2000).
13. *Le Opere di Galileo Galilei*. vol. XI.
14. K. v. Gebler, *Galileo Galilei and the Roman Curia* (London, 1879).

15. Ibid.
16. Ibid.
17. Ibid.
18. Ibid.
19. Ibid.
20. Ibid.
21. M. Falorni, "The Discovery of the Great Red Spot of Jupiter," *Journal of the British Astronomical Association* 97, no.4 (1987): 215–219.
22. J. H. Rogers, *The Giant Planet Jupiter* (Cambridge, UK, 1995).
23. v. Gebler, *Galileo Galilei and the Roman Curia*.
24. Ibid.
25. A. Koestler, *The Sleepwalkers* (London, 1959).
26. Ibid.
27. M. White, *Galileo: Antichrist* (London, 2007).
28. Ibid.
29. G. Galilei, *Against the Calumnies and Impostures of Balthasar Capra, etc* (Venice, 1607).
30. Koestler, *The Sleepwalkers*.
31. G. Galilei, *Letters on Sunspots* (Venice, 1612).
32. Koestler, *The Sleepwalkers*.
33. Ibid.
34. Drake, *Discoveries and Opinions of Galileo*.
35. *Le Opere di Galileo Galilei* vol. XX (1972).
36. G. Galilei, *History and Demonstrations about Sunspots and their Properties,* trans. Stillman Drake (Venice, 1613).
37. *Le Opere di Galileo Galilei* vol. XI.
38. Ibid.
39. Letter to Cardinal Allesandro d'Este, January 20, 1616, trans. Giorgio de Santillana.
40. D. Goldsmith, *Einstein's Greatest Blunder?: The Cosomological Constant and Other Fudge Factors in the Physics of the Universe* (Harvard, 1995).
41. Letter to Cardinal Allesandro d'Este.
42. Ibid.
43. Ibid.
44. Ibid.
45. *Le Opere di Galileo Galilei* vol. XI.
46. G. de Santillana, *The Crime of Galileo* (Chicago, 1955).
47. Ibid.
48. Ibid.
49. Ibid.
50. Ibid.
51. Ibid.
52. T. Kuhn, *The Copernican Revolution: Planetary Astronomy in the Development of Western Thought* (Harvard, 1957).
53. de Santillana, *The Crime of Galileo*.

54. Ibid.
55. Ibid.
56. *Le Opere di Galileo Galilei*, vol. XVII (1972).

Chapter 9

1. R. Davies, *The Darwin Conspiracy* (London, 2008).
2. L. Price, *Dialogues of Alfred North Whitehead* (New York, 1956).
3. Darwin's letter to Lyell, June 3, 1858, *Encyclopaedia Britannica*, 15th ed.
4. J. Marchant, *Alfred Russel Wallace: Letters and Reminiscences,* 2 vols., (London, 1916).
5. G. DeBeer, *Charles Darwin, Evolution by Natural Selection* (New York, 1964).
6. C. D. Darlington, *Darwin's Place in History* (Oxford, 1959).
7. L. Eisley, *Darwin and the Mysterious Mr X: New Light on the Evolutionists* (London, 1979).
8. Ibid.
9. Davies, *The Darwin Conspiracy.*
10. Ibid.
11. Ibid.
12. Ibid.
13. Ibid.
14. Ibid.
15. Ibid.
16. Ibid.
17. Ibid.
18. C. Darwin, *The Descent of Man, and Selection in Relation to Sex* 1st ed. (London, 1871).
19. M. Fichman, *An Elusive Victorian: The Evolution of Alfred Russel Wallace* (Chicago, 2004).
20. Ibid.
21. M. Ridley, *How to Read Darwin* (New York, 2005).

Chapter 10

1. Quoted in Robert Miklitsch, "Performing Difference: Brecht, Galileo, and the Regime of Quotations," *Journal of Dramatic Theory and Criticism* (Fall 1991), University of Kansas.
2. Ibid.
3. B. Brecht, *Life of Galileo,* trans. Desmond I. Vesey (London, 1960).
4. Ibid.
5. B. Brecht, "Life of Galileo," trans. Charles Laughton, in *Works of Bertolt Brecht,* ed. Eric Bentley (New York, 1952).
6. Ibid.

7. Ibid.
8. Ibid.
9. Ibid.
10. Brecht, *Life of Galileo*.
11. Ibid.
12. Ibid.
13. Ibid.
14. Ibid.
15. M. White, *Galileo: Antichrist* (London, 2007).
16. A. Koestler, *The Sleepwalkers* (London, 1959).
17. Ibid.
18. White, *Galileo: Antichrist*.
19. Thanks to Adam Walton, BBC journalist, private correspondence.
20. Brecht, *Life of Galileo*.
21. Michael Brooks, "The Years of Thinking Dangerously," *New Scientist* (December 2008).
22. Ibid.
23. Ibid.
24. Ibid.
25. Ibid.
26. Ibid.
27. Ibid.
28. Ibid.
29. Ibid.
30. J. Gregory and S. Miller, *Science in Public: Communication, Culture and Credibility* (Cambridge, MA, 1998).
31. C. Darwin, *The Origin of Species by Means of Natural Selection* (New York, 1872).

INDEX